ALTERNATIVE SCI

ALTERNATIVE SCIENCES

Creativity and Authenticity in Two Indian Scientists

Second edition

ASHIS NANDY

UNIVERSITY PRESS

OXFORD
UNIVERSITY PRESS

YMCA Library Building, Jai Singh Road, New Delhi 110 001

Oxford University Press is a department of the University of Oxford. It furthers the University's objective of excellence in research, scholarship, and education by publishing worldwide in

Oxford New York

Athens Auckland Bangkok Bogota Buenos Aires Cape Town Chennai Dar es Salaam Delhi Florence Hong Kong Istanbul Karachi Kolkata Kuala Lumpur Madrid Melbourne Mexico City Mumbai Nairobi Paris São Paolo Shanghai Singapore Taipei Tokyo Toronto Warsaw

with associated companies in Berlin Ibadan

Oxford is a registered trade mark of Oxford University Press in the UK and in certain other countries

Published in India
By Oxford University Press, New Delhi

First published by Allied Publishers 1980
Second edition, Oxford University Press 1995
Oxford India Paperbacks 2001

ISBN 019 565 5281

Typeset by All India Press, Kennedy Nagar, Pondicherry
Printed in India by Rashtriya Printers, New Delhi 110 032
Published by Manzar Khan, Oxford University Press
YMCA Library Building, Jai Singh Road, New Delhi 110 001

Contents

To the Ramanujans
who walk the dusty roads of India undiscovered
and the Boses
who almost make it but never do

Preface to the Second Edition

Twelve years is a long time in a person's life; it is an even longer time in the life of a book; and nothing can be as startlingly educative as reading one's own work after such a gap. The earlier introduction to *Alternative Sciences* reflects that discovery. It was written, as I make clear in it, some years after the work on the book was completed and gave me a chance to have a second, more dispassionate look at the whole endeavour. It was in fact an author's attempt to mark out a possible baseline for criticism of his own book. I propose to use this preface to strengthen that process.

I am still fond of *Alternative Sciences*, and I still wistfully remember the leisurely pace at which I worked on the two life histories included in the book. It was after all my first independent work of scholarship. However, when I wrote the book, my primary emphasis was on the social psychology of creativity. So I accepted, even if grudgingly, the standard two-fold division of science into its text and context. Over the years I have given up this two-fold division, and come to believe that there is no Orwellian thought-police guarding the borders between the contents and contexts of modern science. The main function of the dichotomy has been to deflect all criticism of science away from the scientist, towards the forces that control the externalities of modern science—some well-known candidates being the military–industrial complex, American hegemony, Stalinism, Oriental despotism, and religious bigotry. This is what I have elsewhere called the principle of split legitimation, influenced partly by the work of my friend J. P. S. Uberoi. According to this principle, only the good that science does need be owned up by scientists, never the evil. The politically powerful play up to the split, for it allows them to bypass the scientists' moral selves while appropriating their professional skills. Few claim that the Weimar Republic played any role in the creation of quantum mechanics or the theory of relativity. Almost everybody believes they were the discoveries of gifted scientists; it was nuclear weaponry that was the doing of necrophylic establishments.

In that sense I now find *Alternative Sciences* not adequately critical. The two scientists I have constructed in its pages are ambivalent towards modern science. They do use traditional cognitive orders as a baseline for social criticisms of modern science, but they both feel uncomfortable with the idea of the plurality of science and the proposition that there can be politics or culture in the content of science. Like the author when he wrote the book, they want to give modern science a richer, more plural context and to bring that context to bear upon the creative process in science.

In other words, I blame myself for not pursuing systematically the alternative frames of knowledge that the two scientists used and leaned against. For instance, in the section on Srinivasa Ramanujan, there is no discussion of the indigenous schools of mathematics which had survived in vestigial form in south India, specially in Kerala, and which shaped his mathematical worldview. Even in the case of Jagadis Chandra Bose, there is insufficient effort to set Bose's vitalistic biophysics within the tradition of Indian science. To those who know my subsequent work, I must issue the warning that this is not a book on knowledge systems; it is a book on the personal contexts of scientific creativity outside the normal habitat of modern science. This has given a touch of incompleteness to the two life stories, especially to my profiles of the two scientists as functioning, creative individuals. Fortunately other scholars have in the meanwhile worked on this part of Ramanujan's heritage and what I left undone in his case has not remained an area of darkness any more.

If I were to write this book today, I should define my responsibility slightly differently. Taking a position with the victims of modern science outside the perimeters of the 'civilized' world, I shall offset myth against science and history. These victims, for good or for ill, tend to speak the language of myths; their rulers and their self-declared emancipators both speak the language of science and of history. My inability to build that into the very framework of my analysis would be a more serious criticism of this book than that new philosophical insights have led to a revaluation of Bose's plant physiology, or that new data have been amassed on Ramanujan's life and work.

Some other opportunities which I did not exploit sufficiently are probably now, sadly enough, no longer open to exploration. I did not fully know what to make of the rich experience of interviewing

Janaki, Ramanujan's greatly-suffering wife. I could sense her powerful presence, her meaningful though sad life after her husband's death, and her instinct for survival. She was an interesting person in herself, not merely as her husband's shadow or her mother-in-law's other. Similarly with John Littlewood and C. P. Snow. My framework did not have much space for their captivating personalities which told me so much about the pre-war culture of science. Even though I sensed that these men bridged, through their self-definitions, different eras and traditions of science, I did not know how to integrate these experiences into my narrative. I had spent hours with all three of them and my encounter with Janaki was particularly moving; but I stuck to the personality of Ramanujan and the environment that shaped it, and did not digress into the trajectories of others who had played a role in his life. Similarly, towards the end of my work on the book, I came to know that Mrs Boshi Sen—her husband was at one time one of Bose's intimates who parted company with his hero as a disappointed and slightly bitter apostate—lived at Almora at the foot of the Himalaya. Reportedly she was a storehouse of information on the later Bose, his style of research, his organizational and interpersonal skills. Alas, I felt I knew that part of the story rather well and did not need to reopen the case I had already studied.

Probably, *Alternative Sciences* would have been a more acceptable book for me today if I had consciously allowed it to be something more than a book on the cultural psychology of scientific creativity.

Yet, despite being unavailable for over five years, this book has continued to enjoy a certain patronage from critics and students of the sociology and culture of science. Some readers have generously continued to look at it as a step towards the identification of a more humane, culturally rooted system of science. Even if this is not so, the book remains to this day one of the few available on the cultural context of individual creativity in science in the non-western world. It is with that awareness that this edition is being issued. I hope it still manages to convey to the reader, however imperfectly, something of the anguish of being a dissenting scientist in this part of the world, with the defence of cultural and intellectual values from the point of view of defeated systems of knowledge, and with the psychological costs of confronting an imperial system of knowledge outside the western world.

I am grateful to my editor at the Oxford University Press who took the initiative of republishing the book and also helped me get rid of the touch of pomposity that often informs an author's first work, to Punam Zutshi who provided editorial help and words of wisdom, and to Bhuwan Chandra for his secretarial services.

Finally, two warnings to the reader that I should have given when the book was first published. Many technical terms used in the following pages do not appear to be so. I have in mind especially clinical terms borrowed from psychoanalytic psychology such as projective system (*à la* Abraham Kardiner) and projection, identification, reality-testing, rationalization, ritual and ritualization, orality, compensation, reparation, reality-testing, regression, cathexis and concretism. Even apparently everyday words such as need and nurture (both used in Henry Murray's sense), self, denial, identity, development (in the sense of personality development), motive and defence have technical meanings. I hope that the sense in which I have used them is clear from the context; nevertheless, the qualification should be borne in mind.

Second, I have not come across any important new work on Bose during the last two decades. The brief chapter on him in Peter Tomkins and Christopher Bird's *The Secret Life of Plants* is primarily a response to the changing attitude to science in the early 1970s. In the case of Ramanujan, however, there is now a charming new biography, Robert Kanigel's *The Man Who Knew Infinity*. There have also been new developments in the study of his work, thanks to the rediscovery of some of his notes.

Delhi, 1993 A. N.

Preface to the First Edition

Psychologists analysing biographical data are twice privileged. They may write bad life histories because they are psychologists and they may produce bad psychology because they are dealing with history. I have brazenly exploited both the privileges in this book. The only mitigating circumstance I can claim is that somebody had to raise the issues raised here. This is one book which was in search of an author. That it found one in me is an irritant which the reader will, regrettably, have to endure.

I have another confession to make. The science of science is today a specialized discipline. It was less so when I started this work some years back. So the reader may find in the following pages a certain naïveté, theoretical as well as methodological. Unfortunately, without rewriting the entire book it was impossible to utilize fully the new developments which have since taken place in social studies of sciences and to remove this naïveté. I am ashamed to say that I have shirked that responsibility. While that makes this book technically a less competent enterprise, I hope it still communicates some flavour of the inner world of two Indian scientists.

To make the task of my reviewers easy, let me briefly state at this point what this book is about. In the following pages I have analysed the life histories of two scientists who at one time seemed to promise new paradigms of science as well as new models of scientific creativity.

The first essay is on Jagadis Chandra Bose (1858–1937) who started his career as a brilliant physicist, changed his discipline to become an even more influential plant physiologist, and died a lapsed scientist and half-forgotten mystic. In his heyday, his admirers such as Albert Einstein, Bernard Shaw, Henri Bergson, Aldous Huxley, and Romain Rolland found in him the personification of a historical civilization which had a more humane concept of science and a more integrated view of the organic and inorganic worlds than the West would offer. Even when he had fallen from grace in the world of science, his compatriots continued to see in him a symbol of Indian science and a pioneer who had Indianized modern science to make it compatible with the culture of an ancient society. The essay traces Bose's science to his early socialization, the distinctive concept of science in his society, and the needs of modern science at his time. It also shows how his professional degeneration reflected the interactive demands of his subject society, his personality, and the apparently universal culture of world science dominated by societies which in the context of India aroused deep feelings of personal inadequacy and a painful search for parity.

The second essay is on Srinivasa Ramanujan (1887–1920), one of the greatest untrained mathematical geniuses ever known. When this was accidentally discovered at the age of twenty-five, he was working as a clerk in Madras on a salary of twenty-five rupees a month. But on his own, helped by a hopelessly out-of-date, second-rate textbook for undergraduates, he had re-made some of the major mathematical discoveries of the previous hundred years and was in many fields far ahead of his contemporaries. The essay analyses the sources of Ramanujan's mathematical thinking, the association between it and his magical–ritualistic concepts of numbers and manipulation of numbers, his uncontaminated orthodoxy and pride in it, and the psychosocial meanings of the bonds he had established with his famous discoverer and collaborator, G. H. Hardy, a man in turn driven by his homosexual needs and his deep ambivalence towards conventional authority systems. In the process, I explore the extent to which Ramanujan's culture and his unexposure to a modern lifestyle spared him the internal conflicts of Bose, the extent to which his orthodoxy ensured his autonomy as a creative mathematician, and how his private meaning of mathematics gave him a valid personal philosophy of science so that he could live and die a functioning mathematician. The essay also includes a brief comparative discussion of the ways in which Bose and Ramanujan mediated the demands of modern science, their own motivational patterns and the traditions of their society. It ends with some theoretical speculations about the common meanings of inspiration and creativity in two men who might have been by-products of the cultural changes in their country and in the world of science.

The first chapter provides an introduction to the two essays, discussing the normative implications of these psychological and historical accounts and the theoretical concerns guiding them.

Many persons and environments have contributed to the writing of this book. I can mention here only the most important of them.

D. M. Bose, Gopal Chandra Bhattacharya, Pulin Bihari Sen, and S. N. Bose helped me with a mass of information about Bose's life and work, and Rajni Kothari, S. K. Mitra, Steven Dedijer, Avery Leiserson, Edward Shils, Lucian Pye, and D. L. Sheth commented on an earlier version of the section on Bose. Lloyd and Susanne Rudolph, Milton Singer, R. S. Das, Nihar Ranjan Ray, and Raymond Owens were some of the first to read my early draft on Bose and to encourage me in the venture. To David Edge and the anonymous referees of *Science Studies* who published a paleolithic version of the section on Bose, I owe many useful editorial suggestions.

Janaki Ramanujan, P. K. Srinivasan, J. E. Littlewood, S. Chandrasekhar, and C. P. Snow supplied me with little-known or unknown details about Ramanujan's life and about the Ramanujan–Hardy relationship. Prakash Desai wrote what virtually amounted to a review article on the Ramanujan section. His suggestions, some of which I have used in this version of the essay, also included a fascinating, comparative evaluation of Bose and Ramanujan from a psychiatrist's point of view. Once again D. L. Sheth made detailed comments, in his distinctive beatific style, on an earlier draft of the section, and C. R. M. Rao, Indukant Shukla, and C. Chintamani went with their fine editorial tooth comb through my Bengalese to turn it into tolerable English.

Section three of the essay on Ramanujan was also published in the *Psychoanalytic Review*. It owes its present shape partly to the useful criticisms of the anonymous reviewer of the journal. The Indian Council of Social Science Research provided, at very short notice, funds which permitted me to examine the Ramanujan Papers at Trinity College, Cambridge, and to interview Janaki Ramanujan, Littlewood, Snow, D. M. Bose, P. B. Sen, and S. N. Bose.

For all this generous help I am immensely grateful. But, above all, it is to the group at the Centre for the Study of Developing Societies that I owe my gratitude. By constantly confuting my position on extra-scientific determinants of science while I was writing the section on Bose and by as vehemently opposing whatever autonomy I granted to science while writing on Ramanujan (because by that time they had partly convinced me by their arguments) they may have further shaken my already-tottering belief in human self-consistency, but they helped me clarify my own concepts and sharpen my analytic tools. Though I have not been able to satisfy them fully in either of their incarnations, they have more than satisfied my need for inntellectual succour.

Delhi

Ashis Nandy

Part One

Introduction

The Alien Insiders

I met Jagadis Chandra Bose accidentally. I had agreed, against my better judgement, to attend one of those typical seminars on psychology and science. It was while rueing this decision that I came upon a book of essays by Bose, republished on the occasion of his birth centenary. Written in impeccable Bengali, and often frankly autobiographical, these essays projected a distinctive concept of science and gave tantalizing clues to the personalized meaning given to science by an Indian scientist. Deeply impressed, I decided to write an essay on Bose. I was then fresh from a stretch of training and research in a psychiatric clinic, and Bose's complex personality with its articulate inner concept of science and interacting vectors of tradition, creativity, and search for new modes of intellectual self-expression captivated me. All that had to be done, it seemed, was to offset these interactions against his psychodynamics to produce a reasonably interesting conference paper. I then had no feel for the cultural and ethical problems of science. And the resulting essay, published elsewhere, reflected this.

But the subject grew on me. Mainly because, while I did something for Bose, he also did something to me. I tried to give him a new historical image which stretched beyond recognition of his virtuous, two-dimensional image, created, as it then seemed to me, by naïve biographers, credulous historians, and hypocritical journalists. Almost without trying, merely by being faithful to my discipline, I allowed the tempestuous scientist to be a little more like his real self and a little less of the play actor which he always was, but infrequently resented being. If my analysis had somehow reached Bose, he would certainly have thrown a tantrum. But perhaps he would also have been happy to have his façade pierced. After all, he had to bear alone all his life the painful, weary conflicts between the traditions of his society and the traditions of science, between a westerness associated with the culture of his country's rulers and a westernness characterizing the ruling culture of modern science, between science as an ideology and science as a gentlemanly hobby, and, above all, between an integrated but loose-ended Bose at peace with the world and himself and a troubled, authoritarian, insecure Bose who even had to overcome mental illness without professional help. (Bose's friend Girindra Shekhar Bose was India's first psychoanalyst, working next door to Bose's home and laboratory. The parallel professional life that the other Bose led, uncontaminated by and unaware of the psychological sufferings of his friend, shows how much one had to live with one's private ghosts even a few decades back.)

Jagadis Chandra, on the other hand, sensitized me to the psychological and cultural predicaments of modern science. Not through the fashionable, glib language of the so-called science of science, but by nearly succeeding and then failing as a creative scientist who hoped to delineate for Indian scientists the outlines of a possible collective identity. The identity he evolved did not work, but the effort was, to say the least, impressive. He suffered greatly as he moved from the confident innovativeness of those hopeful days when it had seemed possible to challenge successfully the western dominance of science, to the insecure, dogmatic, ideological postures that accompanied his humiliating awareness that he had failed to break this dominance, and ultimately to the even more humiliating refusal to recognize his failure. But his sufferings only showed me how much insight there was in his perception that the western culture of science was not immutable, and that the Indian search for autonomy *in* science was, actually, a variation on the age-old struggle for the autonomy *of* science.

But what is commonplace today was outside the ambit of scientific self-consciousness then. And Bose never felt that his innovations in scientific culture had any chance of survival without his success in formal science to back it up. As I moved around with him in his world, it soon became obvious that this was the most constricting of the blinkers he wore. He was not willing to accept himself as merely a creative philosopher of science. He had also to succeed as a scientist. Yet, certainly his failure as a scientist would have been forgivable. It would in fact have been surprising if he—working in isolation in a poor, colonial country—had won his battle against a science which was increasingly becoming an organized, capital-intensive, group effort—a 'big science', as many have recently begun to call it. Apart from this, it needed a man of Kepler's and Newton's abilities to end the western dominance of modern science by altering both its content and its metaphysic. Bose's gifts—his self-deceiving countrymen and western admirers notwithstanding—were much less formidable.

It was his failure to accept that he had ceased to be a creative scientist which was to have dangerous consequences for the culture of science he wanted to build. He became more authoritarian, his aggression became free-floating, and his allegiance to a science which had failed him and his society also got eroded. In middle age, he allegedly began manipulating his experimental results, advancing false claims, and encouraging exaggerated statements by influential friends and journalists about his achievements.

What were the signposts on the way from his early successes to later failures? How correctly did he read them? How responsible was his internal equipment for what he became or could not become and how much responsibility did his society and the nature of science bear? These were some of the questions that prompted me to look for a contemporary of Bose who would represent the other end of the spectrum as far as Indian responses to modern science were concerned. My choice of Srinivasa Ramanujan as a second witness was, therefore, a more deliberate and reasoned one. It was, however, a statement of G. H. Hardy that finally convinced me that Ramanujan was the scientist one should counterpoise against Bose. In his book on the Indian mathematician, Hardy says at one place that the 'solitary Hindu clerk' had pitted his brains against more than a century's accumulated knowledge of the West and that he was bound to lose. This reminded me of Bose's life-long pursuit of a grand defeat that, by its sheer grandeur, would outdo any success he could achieve. What Bose sought so consciously and in his studied way, the Tamil Brahmin was seeking in his inarticulate, introversive, unselfconscious manner.

But Ramanujan proved to be a more intractable subject than Bose. He had the kind of modesty which the progeny of a defeated civilization rarely show. Absolutely certain about the correctness of his concept of science, he left his professional colleagues in no doubt of his certainty. He never felt called upon to justify himself, except to himself. He articulated his self-confidence in a low key, without making an issue of his philosophy of science, and without the conspicuous self-righteousness which is often the hall-mark of Indian piety. He left no autobiographical writings, few letters which did not start or end with something on mathematics, and even fewer friends who could understand both his mathematics and lifestyle. (The last is the most difficult to forgive. Not leaving an autobiography is understandable, particularly if one happens to die at the age of thirty-two. One could be impersonal or lethargic about writing letters. But not leaving behind some friends as informants for future psychologists can only be called thoughtless.)

Yet, it was not so much the scarcity of data as the wrong kind of data from friends on Ramanujan which were for me the reddest of herrings. His modernist admirers wanted him to be modern; the traditionalist ones wanted him to be magical. Between the two sets of friends, Ramanujan managed to have two implicit biographies, two career lines, and two interpretations of science.

It was while trying to reconcile these two Ramanujans that I became aware of the typical orientations to history and biography obtaining in the Indian society. And, what at first had appeared to be wrong facts, mystification, and convenient amnesia, came to acquire a new meaning as well as validity. Brought up on the orientalist belief that Indian culture was ahistorical, I had failed to notice the three distinctive attitudes to the past which characterized India's popular 'histories' and sometimes even its modern historical scholarship. Each of these attitudes, as the following discussion will make obvious, is an ideal-type by itself. But together they provide an analytic framework within which can be located the peculiar Indian attitudes to past times and persons.

The first attitude is that of the traditional minstrels or *charans*. They were wandering individuals or small nomadic troupes who sang of past events and men—giving meaning to the present by projecting its rough realities into a mythologized past so that the present became more 'livable'. Moving from village to village, singing of things the villagers would never live with but were expected to live by, they made history a folk art—a shared fantasy, if you like. While providing a capsuled world image and organized ethical criteria to the laity, they built a defensive shield which consolidated the culture through constant affirmation and renewal of its psycho-philosophical base.

The second attitude is that of the Brahminic *barots* who were genealogists and chroniclers. Using the legitimacy traditionally given to 'pure data' uncoloured by emotions, they communicated a sense of formal generational continuity by ignoring flesh-and-blood individuals. No sanctity, however, attached to the totality of facts. It was through a selective presentation of facts, including a huge load of trivia, that the genealogist made his point. His very affectlessness came in handy in this; he used it to reify all human relationships and make history an impersonal, dehumanized, abstract, tireless 'mathematics'.

Thirdly, there were the court historians and their humbler versions, the *bhats*, who sang praises of their royal employers. Such an attitude, however, need not be confined only to court historians and their royal patrons; others too created mythical, larger-than-life subjects out of the mortals who happened to be the object of their interest. Everyone, including the subjects of praise, knew this to be a game and nobody took the content of the praise seriously. It was only the form of praise that was ritually important.

All three attitudes are deeply embedded in an orientation to the past which reduces—or elevates—each social reality to a psychologically significant myth. The aim is to wipe out the historical reality altogether or supplant it by the structural realities of the mythical. It is then from the second set of realities that the 'secondary' historicized facts of a person's life are deduced.

The three attitudes have given the traditional Indian concept of the past a certain distinctiveness. The first attitude subordinates the historical reality of past individuals and individual events to the process of cultural continuity and cultural renewal. While it destroys the individuality of historical persons and events, it simultaneously invests them with the individuality that could only be given to them by their civilization. So that the present can always see a continuity between exemplary persons from the past and the perennial concerns of the society. This is how Indian society had taken care of Benedetto Croce's and R. G. Collingwood's theories of history.

Ramanujan, being a more traditional man from a more traditional background, gives one greater scope to explore this cultural individuality and its relationship with the metaphysic of science. Bose, a more modern man from a more modern subculture, seems to survive his biographers as a historical individual. In other words, contrary to what I had expected, while Ramanujan gave me new access to his culture through his life history, Bose's culture guided me to another approach to Bose's life. It is all a matter of emphasis, but this difference is reflected in the two following essays.

The second attitude to history reduces men and events to data and statistics, so as to neutralize any emotion that may be associated with them. The aim is to dissociate affect from cognition and indirectly express the former through a manifestly desiccated recital of figures and events. That is why, in traditional Indian historiography, the data produced and the statistics used are often unique. A king is mentioned as having sixty thousand children, and the heavens are mentioned as being inhabited by three hundred and thirty million gods, not only to make the point that the king is potent and the gods are many, but also to wipe out what many would consider the real data, and obviate any possibility of verification or empirical treatment. If you have sixty thousand children, no one cares to ask their names or whereabouts, not even the single-minded orientalist; and no sacred text need list the three hundred and thirty million gods, even for the priest. In fact, you are not expected to take these figures seriously. You can only counterpoise against these communications others more potent. In other words, in this type of historiography data are important only so far as they relate to the overall logic and the cultural symbols that must be communicated.

There is a remarkable isomorphism between this attitude and the scientist's attitude to the history of science.

> The more historical detail, whether of science's present or its past, or more responsibility to the historical details that are presented, could only give artificial status to human idiosyncrasy, error and confusion.... The depreciation of historical fact is deeply and probably functionally ingrained in the ideology of the scientific profession, the same profession that places the highest of all values upon factual details of other sorts.[1]

So, there were all the details of dates, degrees, and awards to Bose and Ramanujan, their genealogies, and the lists of their publications. But, in the case of Ramanujan, only one person mentioned his attempted suicide and most remained silent about his conflicted relationship with his family. In the case of Bose, none would mention the details of his 'nervous breakdown', the insomnia that dogged his adult life, or the stutter that he broke into when angry. Again, very few cared about the personal relationships of these men, or the role religion, as opposed to magic, played in their

[1] Thomas Kuhn, *The Structure of Scientific Revolution* (Chicago: University of Chicago, 1962), p. 137.

lives. Ramanujan, in this respect, was more unfortunate. His Tamil Brahmin culture was more foolproof, more confident of its answers to the enigmas of inspiration and creativity, and more contemptuous of whatever did not fit into the frame of these age-old answers. Most of his Indian biographies are, analytically, carbon copies of one another. His English friends, on the other hand, wanted him to be an ahistorical, agnostic positivist—a one-dimensional scientist, trying singlehandedly to explode the myth of the inscrutable orient.[2] Bose's *babu* environment, no less conformist in biographies, was at least more curious about the totalities of its great men.

The third attitude to history makes for the delightful exaggerations and absurd eulogies in which most Indian biographies of Ramanujan abound. For these biographers, he was a scientist as great as Newton, a man as saintly as Gandhi, and a mystic as awe-inspiring as Aurobindo. Similarly, Bose for most of his biographers, was Einstein, Houdini, and an ancient Indian *rishi* all rolled into one. Partly this is a studied style which attempts to treat individuals as symbols rather than as historical figures. Epic culture, we know, survives on such symbols. Partly this can be traced to the long colonial status of the society. Biographers are happy to stretch the reasonable statement that Ramanujan might have been the greatest mathematician of all times but for his poverty and lack of formal education. Or that Bose could have been one of the most creative plant physiologists but for his acute sensitivity to the western dominance of science and his professional isolation in India. Still suffering from some of those very handicaps which distorted the career lines of Ramanujan and Bose, these biographers like to see their eminent *alter egos* as perfect men who, in the face of heavy odds, ultimately managed to actualize their full potentialities. They fail to grasp that through such distortions they try to do the impossible, namely, simultaneously overvalue and deny the existence of the twin enemies of scientific pioneering in India, economic deprivation and political subjugation. On the one hand, they stress the

[2] India's modern scientists of course fall for the imported amnesia. I remember my long unresolved arguments with two scientists while seeking foundation support for the work on Ramanujan. Both tried to convince me that any psychological exploration into Ramanujan's personal life would be futile because Ramanujan's life was a 'short, straight line' and there was nothing to it except his mathematical career. They did not know that Littlewood, Ramanujan's friend and teacher, had made a better argument out of this. Whether Ramanujan had a private life or not, Littlewood had told an acquaintance after being interviewed by me, it was beyond the reach of a psychologist. For psychologists, he felt, could never understand mathematicians.

crushing poverty within which Ramanujan worked in his youth; on the other, they emphasize his easy victory over all the handicaps which poverty set for him. On the one hand, they mention Bose's life-long problem with the colonial situation; on the other, they presume that the impact of colonialism on Bose's science and personality was skin-deep.

In the following narratives I have tried to show that Ramanujan's poverty and his inability to get a good mathematical education were more killing than his biographers seem to realize. And colonialism subverted Bose's creativity more successfully than the nationalists seem to suspect. Poverty and colonial status become totally corroding when they enter a person's or group's self-definition and become psychological forces. In the case of Bose and Ramanujan, this internalization did take place and they had to fight their demons within as well as without.

The third attitude to history is legitimized by the indigenous theory of personality development implicit in many Indian biographies. This theory is a unique blend of inverted Freudianism and folk genetics. It presumes that the early within-family relationships of great men are not merely idyllic, but that the great qualities of the great invariably develop out of their relationships with the best of all possible parents and siblings. It is as if the Indian culture accepted the importance of early child-rearing and the primacy of a person's object relations and denied only their total content.[3] Bose and Ramanujan were two of the worst sufferers from this syndrome, partly because they themselves held the same attitude and shedding it would have meant shedding their psychological defences. Fortunately both men, while often uncommunicative in their garrulity, were always eloquent in their silence.

It was not only the silence of my protagonists' witnesses that was eloquent; that of their admirers was equally so. I never felt cheated when one of them prefaced his statement with the request that I keep to myself what he was going to say. When they narrated in hushed tones—as if speaking of some saintly pecadillo—an anecdote or a reminiscence, I had almost always at least the outlines, if not a new micro-interpretation, of an old myth. I valued

[3] There is a contradiction between this attitude and the Indian epic tradition which, like the Homeric and some other epic traditions, has an amoral, tragic sweep which confidently—almost carelessly—takes into account a person's life in its entirety. I shall avoid, for the moment, a discussion of this contradiction.

this immensely. These lapses into Indianness were the bricks of my analysis, because I was as much in search of the real lives of the two men as after the myths generated by the interactions between their lives and the large-scale psychological and historical changes taking place in their society. I believe that these two types of 'truths' and their interchanges will ultimately prove to be invaluable source material for a social psychology of Indian science.

I have been silent till now about the creative products of Bose and Ramanujan as sources of biographical insight. Presumably this silence, too, has been significant. I am neither a plant physiologist nor a pure mathematician. It was only an accident of personal history that I was able to read and understand some of Bose's scientific writings first hand. I did try to take a guided tour in the more trodden sectors of Ramanujan's as-yet-partly-charted mathematical world, but it was beyond me. Thus in the case of Ramanujan entirely, and in that of Bose partially, I had to depend on the interpretations and estimates of others. And because these interpretations and estimations were not inspired, understandably and legitimately, by the same concerns as mine, my understanding of both as scientists has remained limited.

Before proceeding further, a word about this 'dangerous' limitation. When I started work on this project, many appreciated my 'courageous' decision to study the scientists. This was their way of saying what the more outspoken of my acquaintances had bluntly asked: how could I dare write on these scientists without fully understanding their sciences? In impolite moments, I responded with a counter-question: how could a scientist hope to write about these men without fully understanding their humanity?

But all that is polemics and a product of the insecurity generated by the so-called hard sciences among social scientists. My real answer is this: the statistical probability approaches zero of someone combining in himself the skills of a psychologist and a cultural anthropologist, specialized training in one or more scientific disciplines (to understand Bose's work one would require a knowledge of sciences as disparate as physics and plant physiology) and interest in the life of scientists. To expect exactly some such combination to emerge in the case of Bose and Ramanujan would in fact be stupid. The chances are that everyone will be deficient in at least one or two of these admirable qualities. An indirect confirmation of this came when I found a number of mathematicians ignorant of Ramanujan's mathematics and a large number of physiologists admitting their ignorance and lack of interest in Bose. If even being a mathematician or a physiologist was not sufficient qualification in this age of specialization, I decided that I had done my best for my historical friends on behalf of posterity by remaining a mere psychologist.

II

I now turn to the other party involved in my encounters with these two Indian scientists: the nature of science as we have come to conceptualize it in this century. This is a notoriously protean subject, which includes today both the cultures and the histories of the natural and social sciences and their growing interlinkages. But the logic of Bose's and Ramanujan's handling of science as a medium of human self-expression lies partly in the political psychology of contemporary science. And you should know my likes and dislikes in this area too.

In the following pages I have avoided regarding science as a social process which pays off in technology and other forms of applied knowledge. Generations of social thinkers have discussed the political and ethical responsibilities of scientists at this plane. The public, too, likes to think of science as a problem-solving technology which can cure some of the world's major ills. Given this wide interest in science as technology, I felt that a few more clichés from me on this issue would not enrich existing knowledge.

I have also avoided approaching science as a system of cumulative, structured thought. Scientists like to view science as something impersonal, objective, and free from the limitations imposed by human consciousness and history. Laymen endorse this view because they find it convenient to believe that there is at least one area of life which deals only with the constants of human knowledge and rationality. Being in many ways a product of the interaction between scientists and the laity, the culture of science, too, promotes this view to impersonalize and dehistoricize science. (Note for instance, the difficulty of associating the disciplinary heroes of science, the critical individuals who build the sciences, with normal science and practising scientists. An impressive majority of astrophysicists today have probably never read even a word of Ptolemy and few would know anything about him beyond his name. If one compares with this the living presence of Thucydides in historical studies or Plato in political philosophy, it becomes obvious that science must seem to the majority of its practitioners a relatively impersonal, ahistorical vocation.) Perhaps this is one of the main attractions of science to creative scientists, who as a group seem uninterested in their human environment and tend to withdraw from interpersonal realities.[4]

The fact that I do not have anything to say about this aspect of science does not mean that I accept the concept of science as a fully cumulative, impersonal mass of knowledge. Unlike technology, science is a speculative activity and part of philosophy. Naturally, it has its cultural and psychological roots. It is the denial of these roots in the name of an impersonal reified science which, I believe, is responsible for many of the ethical problems of the contemporary culture of science and the limitations of the scientist as a human being. In the following essays too, without going into the content of Bose's and Ramanujan's science, I have tried to show how the abstract, reified structure of scientific knowledge can often act as a defensive shield against both the outside world and the alien aspects of the scientist's own self.

I prefer not to press this point to the bitter end, but it is interesting that many Indian scientists and biographers of scientists whom I consulted while writing this book seemed particularly eager to uphold the image of science as a bloodless affair. Perhaps the image of an affectless, pure science 'clicks' with some vague search for impersonal 'rational' knowledge in India's modernized literati, and particularly with the Indian scientists' search for a viable community identity. Perhaps, as Thomas Kuhn has said, this is part of the world culture of science and the search for a pure, scientific science is a characteristic of scientists as a group. But one thing seems certain: the search does somehow endorse the Brahminic concepts of uncontaminated knowledge and purity of vocation. I remember S. Chandrasekhar telling me that C. V. Raman once administered him a patriarchal rebuke for referring in a speech to Ramanujan's attempted suicide. The rebuke still rankled in his mind five thousand miles away in Chicago, and he gave me a laboured justification for this lapse. I did not then push the point that Raman, in this instance, was articulating a deeply-held belief of both the scientific estate and the Brahminic culture that by objectifying and impersonalizing knowledge and by dehistoricizing

[4] For instance Anne Roe, *The Making of a Scientist* (New York: Dodd, Mead, 1952).

the producers of knowledge one could argue away the imperfect realities of living persons and human history from the world of knowledge. It was a case of perfect fit.

To come to the point, my emphasis in the following pages has been on one particular aspect of science: its use as a creative process which allows the scientist and, through him, his society and times to impose a particular meaning on personal and social existence.

It is only recently that psychological and sociological studies have begun to yield some insigh s into those typical personality assets and the dynamic integration that are necessary for the creative functioning of a scientist. As time passes, one should be able to differentiate between the characteristics specifically associated with scientific functioning and scientific creativity and those predicting intellectual functioning or creativity in general. Meanwhile, in the absence of specific theories, one is forced to work on the basis of ideographic formulations which are only indirectly applicable to other similar cases.

My method was no different. I wanted to explain two persons rather than two sciences, this apologia notwithstanding. Even though my interest in the nature of science has deepened over the years, my implicit dependent phenomena were two real-life scientists rather than the developmental history of Indian or world science. I have tried to understand the latter to the extent it was relevant to my understanding of the former. But by training and exposure, I am relegated—and committed—to the study of scientists rather than sciences.

It is this specific morphogenic concern with individual psychology that made me conscious of science as a culture, with shared concepts of scientific breakthrough, creativity and originality, shared standards of verification, replication, and validation, and shared access to certain anxieties and conflicts associated with scientific functioning. It is this subjective world which makes science the peculiar institution that it is, and has given it the creative form that it has. For instance, when I closely examined the contours of euphoria in the leaders of the scientific estate—Einstein, Raleigh, Haberlandt, Thompson included—once they caught the contagion of Bose's vitalistic science, I found striking similarities between the so-called primitive or traditional science and the concepts of science lurking in the deeper layers of the modern scientific personality. Bergson, Shaw, Einstein, and Huxley—so captivated by Bose's pacifist and ethical science—were certainly not innocent admirers of Eastern mysticism, nor were they innocent of the philosophical basis of modern science. Only, all of them shared certain common anxieties about the nature of modern science and the fate of the technology-based western civilization. In Bose's identity they found elements of the new scientist that the new philosophy of science was looking for.

Thus, we arrive, via a long detour, at a predictable set of questions: Were the older concepts of science that primitive? Or are we, in ignoring them, ignoring major concerns of human destiny? Were Bergson, Einstein, Shaw, and Huxley after all right in their instinctive acceptance of some aspects of Bose's science? True, Bose's Vedantic concepts of holistic universalism and pacifism, to give an example, were defensive and ritualistic. But is that all one can say about them? Culturally, which is the more stupendous effort: to be able to allow a few persons to choose deliberately and actualize the values of integrated knowledge, universalism, and peaceful science, or to be able to turn these into ritually and magically tinted desiderata outside the compass of personal choices? Into, in other words, the institutionalized and internalized parts of a way of life? In the 1970s one cannot be too sure.

Lastly, while there is hardly much debate any longer about the existence of non-scientific determinants of scientific culture, very little intellectual effort has been expended on the process that determines the worldview and ideology of individual scientists. For example, Marxist theory was one of the first to explore the social context of the culture of science, but it has little to say about the actual process through which the social realities come to be reflected in the scientist's personality and work. As a result, when it comes to the individual scientist, orthodox Marxism makes a monstrosity of Marx's humanist interpretation of the social content of knowledge. It becomes a conspiracy theory which relates the material losses and gains of the scientist to his work on a one-to-one basis. Even the Marxist concept of false consciousness becomes for many a means of accusation rather than a technique of analysis. Again, psychoanalysis, which has given us in modern times the most influential theoretical frame for analysing the conversion of social realities into personality vectors, tends to conceive of these realities in static and narrow terms once it goes beyond the immediate interpersonal world of the individual. Its orientation to the larger social forces acting on the individual is like that of a doctor who has dealt all his life with the therapeutics of cholera and is confused when suddenly confronted with an interpretation of the disease which uses poverty as an explanatory variable. The doctor's profession is, by any standard, a legitimate one, and he knows the importance of poverty in the spread of cholera. But he knows it as a technologist to whom the patient's social condition is part of the environment of therapy, not as a scientist to whom the patient's social condition is part of his therapeutic system or scientific theory.

In my encounters with these two scientists, it is on such exchanges between individual creativity and social realities that I have concentrated. And it is the stress on these exchanges that has made the following essays psychological studies, rather than histories of science.

Admittedly, in developing these concerns, I have benefited from the growing self-reassessment of social scientists. Social sciences are called 'developing sciences' and this is as pejorative an expression as 'developing societies'. As a Third-World social scientist, working with the Third World of human knowledge, which is neither the fish of humanities nor the fowl of science, I perhaps could not avoid certain issues of dominance, control, and freedom in science. As everyone knows, living in the Third World is bad breeding; becoming a Third-World social scientist is even worse. The former is like the original sin, the latter like acquired viciousness.

Thus, this exercise of looking into Bose and Ramanujan also constituted an introspective re-examination of the psychological and cultural roots of an Indian interpreter of the world, who explores the outer realities of nature and society from the vantage point of a developing science, with his slanted individual and cultural capacities, with his own attitude to authorities, orders, and pecking orders, and his own concepts of dissent and conformity. I can assure my non-Indian readers that the view from this side of the barrier is quite different.

However, I must give two warnings to those who are always tempted to analyse such self-consciousness to its vivisected death. First, this is not an attempt to support an alternative Indian model of science (as will be obvious from what I have to say about the Indian science of my subjects). I have tried to draw the right lessons from a number of Third-World analyses of social and scientific behaviour which have paid their homage to the West by striving to be the exact reverse of what a hypothetical model of western analysis is. In such cases, even in dissent, the referent is the Occident. Such dissent can be vociferous, but is predefined in such a way that it is always controllable, always on leash, and always a valued ornament of the established authority system of the world of knowledge.

Second, I am not providing a defence for a worldview for those to whom an alien culture, such as the ancient Indian or the modern Chinese, becomes important on ideological grounds. They see in such a culture not merely an alternative civilization, but also a negation of the dominant culture in their own societies. To cater to the need for such an alternative culture, I feel, is to succumb to the existing model of predictability. Take, for instance, the case of Maoist China which for a while had become a powerful symbol of dissent for many in the West. Historically sharing some of the western values of public life and orientations to politics and statehood, Maoist China rejected many of the western forms of politics, economics, and social organization. Its attraction for some sectors of the West was, therefore, understandable. Even the idea of the inscrutable Oriental contributed in the West to the Chinese mystique by hiding the common values of the cultures involved. Thus, China's dissent seemed to many to be, recognizably and comprehensively, a dissent.

Fortunately, in the case of contemporary India, such an inverse relationship between an 'Indian science' and its western counterpart is difficult to establish. The Indian 'alternative', for even the most ardent alternative-seeker, is impossibly unmanageable. It not only seems a half-dissent, it also seems inefficient, chaotic, abstruse, amorphous, and unsure of itself. Its capacity to become a dedicated opponent or even a counterplayer of any other culture is, at a level, much poorer than that of China. And, unlike China, the dominant Indian concepts of public life and collective effort have little in common with those of the dominant societies of the world. It has, therefore, very little to offer to Bertrand Russells, at one end, and Ché Guevaras at the other, men whose conformity to western values was nearly total, and the psychological denial of this fact even more so.

India's 'dissent', in this limited and peculiar sense, is less controllable. That is why Ramanujan's well-meaning English friends tried to make him predictable by remembering him as a respectable, donnish, English dissenter. But he was too much of an authentic Indian alien within the culture of science to be the surrogate of an alienated westerner. And the anxieties he generated, I am constrained to believe, were deeper than the anxieties he lived with. Bose in his innocence, on the other hand, conformed more to the style of dissent acceptable to the West. Europe understood him, and he became the rage for a while. Only, it so happened that those who were lionizing him soon found other trendy things to do. The shrewd Tamil Brahmin was a perfect match for anyone who might have wanted to make him a symbol of protest. Just when the West thought it had captured him, it opened its fist to find that he had slipped through its fingers.

Part Two

Defiance and Conformity in Science

The World of Jagadis Chandra Bose

The image of an omnipotent technology in the last two centuries has blurred the dividing lines between the products of modern science and the formal structure of scientific rationality. As a result, the mystique of science as control and power, created primarily by the Industrial Revolution but also by the certitudes that the doctrines of empiricism and positivism have sired, attaches now also to the content of scientific knowledge. In the name of the autonomy of science, this mystique has encouraged many to assume science to be a distinctive set of ideas, free from the cultural and psychological compulsions operating in other spheres of life. In the form of a lay philosophy of science which distinguishes between science and the scientist, this mystique has helped banish the person from his work and, instead of granting autonomy to the scientists, has granted it to the ideational and material products of science.

Fortunately, the Enlightenment's concept of science, from which the dominant culture of science draws sustenance, is changing and, today, the more sensitive western scientists find it impossible to accept it uncritically. But meanwhile the contagion of the vision of a fully autonomous science has spread, especially in those parts of the world that have begun to seek redemption through the science of development and, as part of that enterprise, in the development of science. There is a widespread tendency among the ruling élites of Asia and Africa to forget that every culture produces its own science as surely as each scientific achievement produces new cultural realities. The price of this forgetfulness can be heavy. As this century has shown, the tendency to see in science a mass of desiccated objective knowledge, untrammelled by human emotions, can hand over entire communities as hostage to the very emotions of which they prefer to remain ignorant. On a lesser scale, it can make the individual scientist captive to the subjectivity he is often tempted to deny and to turn his work into a constricted, defensive manoeuvre.

These are truisms. But one must remember that in communities where science carries the full burden of social hopes and individual aspirations, they remain potent threats. By conceding the cultural and psychological determination of science, in places and times when both cultural and personal systems are in flux, one exposes uncertainties that can lead to crippling anxiety and to a self-examination that can be particularly painful. Yet, when such self-confrontation takes place, it strengthens a society's capacity to use science, in its broadest sense, to creatively redefine the society's concepts of human destiny and fate.

It is in this respect that the insights of the psychology of creativity and cultural psychology have become relevant to the study of science and scientists. They help us to understand the conflicts within ancient cultures which are trying to actualize new self-definitions that would have a place for the modern scientist and, in the process, contribute something of their versions of the eternal verities to the over-professionalized, bureaucratic, technologistic culture of modern science.

A cultural psychology of scientific creativity also allows one to probe the creativity of individual scientists as a link between cultural and individual needs. Such a psychology may not reveal the entire psychosocial landscape within which science and scientists operate, but it can identify for us the type of data that has to be marshalled for such a survey. Methodologically, this has been facilitated by the shift in emphasis in psychology itself. Today the roots of creative functioning are being sought less in the absolutes of human psychosexuality and more in the individual's attempts to cope with society's prototypes within himself. With growing awareness of the cultural and historical roots of the self, the discipline of psychology has been trying to rediscover creative imagination as an interpersonal as well as psychosexual experience. It has moved away from its earlier easy emphasis on the continuities between creativity and unconscious fantasy life, and between creativity and performance on ready-made psychological tests. Instead, one can sense an uncertain shift in the discipline towards a more ethical and cross-culturally sensitive treatment of the cultural and historical processes epitomized in all creativity. As this new emphasis on an intervening, yet culturally and historically rooted, individual relegates to history the shadowy psychological man, the discipline of psychology has been forced to recognize the social content of those personal capacities it chooses to call 'scientific creativity'.

Nowhere does this social determination of scientific creativity create so much anxiety and confusion as in the societies self-consciously trying to draw up new blueprints for their futures. In such societies, science becomes a battleground where the society's new ambitions confront its 'backlogs', and the scientist becomes a microcosm where the community's adaptive capacities challenge the creativity of the individual. In the process, sometimes science itself is distorted and some scientists are destroyed. The society, too, may pay a heavy price, swinging between the extremes of total acceptance of exogenous models of science in society and a doomed search for absolute autonomy in the area of knowledge.

However, broken glass can sometimes act as a prism. The deviant cultures of science in such societies too can, through a process of refraction, give an altogether different perspective on world science, seen as an identifiable psychosocial process.

Nineteenth-century India provides a good example of such a situation. It had a well-developed indigenous scientific tradition, an elaborate entrenched theory of life, and some explicit and implicit rules that related traditional science to the traditional lifestyle. Defying the fit, a handful of men opted for modern science which, though overtly universal, had come to acquire an essentially western culture over the previous three hundred years. Some of them made this imported science their profession, others a rallying call, and still others a symbol of dissent. But in each case they had to opt for a protracted conflict between their Indianness and their professional self and they had to fight the problem of forging, out of these two components, a new oneness that would make sense of their society and to them.

The conflict was sharpened in a colonial society by the western associations of modern science—associations which were bound to make science a symbol of western intrusion. The first generation of modern scientists in India had three possible techniques of coping with the intrusion. The first was to separate the culture and the content of science and, then, fight for pluralizing the existing culture of science in such a way that it would accommodate the Indian worldview. The second was by dismantling the dominant culture of science and replacing it by a new culture more congruent with Indian values. Neither was truly practicable at a time when the difference between science and the culture of science was not a part of common sense in modern India. The first generation of Indian scientists, therefore, opted for a third device. Most of them spent their professional careers trying to build an entirely new Indian structure of science. Some gave up the task half way, finding it too onerous; they preferred to become political activists, institution-builders, or academic bureaucrats and resigned from science, if not formally, at least *de facto*. Others, a smaller group, stuck to their guns and fought a losing battle against the formidable edifice of modern science. All the responses were consistent with the logic of a colonial situation, and one must judge for oneself which was the more tragic dead end.

All three techniques of survival allow us to investigate the limits to which acculturation of a transplanted science can be taken, the extent to which elements of tradition could become functional for modern science, the nature of the legitimacy a modern scientist could hope to shore up in a traditional society where not all the social subsystems were under equal pressure to change, and the process through which cultural and personal realities become the anchors for scientific imagination and creativity. At another plane these historical attempts at transplanting knowledge give us a chance to speculate about the nature of the interdependence of the structure and culture of science.

We shall explore these problems through the life history of Jagadis Chandra Bose (1858–1937), who contributed handsomely to the development of a professional identity which, for a time, seemed capable of mediating between the needs of Indian scientists and the demands on them. His undoubted success on many fronts suggests that his creative imagination could sum up not only the drama of his personal life, but the cultural crisis of his people. A highly successful physicist and botanist, Bose gave a special Indian perspective to world science, and was also one of the first modern scientists to do interdisciplinary research in his field. He also worked out a philosophy of science which anticipated some of the major themes in the contemporary philosophy of science. Simultaneously, he was a savant and a missionary-scientist for many in the West, and a national hero in India.

Yet much of this was to prove ephemeral. Even in his life time, modern physics and botany started overtaking Bose's 'Indian science'; today, within four decades of his death, his idiom sounds flat and out-of-date even in his own country. Though the memory of his achievements survives among important sections of the Indian intelligentsia and popular versions of his life offer a role model to young Indian scientists, his scientific work has already been stripped of its glamour, and his concept of Indian science only marginally enthuses professional scientists. New currents of social change have apparently thrown up newer self-definitions in the Indian scientific community.

Nevertheless, while the struggle to find an endurable professional identity continues amongst Indian scientists, and while Bose's brain-children ambitiously continue to help science rewrite the life-plan of all Indians, his relevance to Indian science persists. The environmental dangers he faced, especially the particularist pressures to which his science ultimately succumbed, are still alive. To the extent these pressures came from his immediate social environment as well as from the larger environment of world science, the forces that made and unmade Bose are relevant not only to the 'underdeveloped' world. They may be relevant even to the technologically ill-developed societies where some of the important questions now dogging science are: How much has science lost by its mechanistic and physicalistic concepts of the universe, how much by its denial of all alternatives to the scientific culture of the industrialized world? How much has the Newtonian idea of a world machine contributed to the ethical predicament of modern science, to its role in fostering human violence, and in violence towards the non-human environment? Can the estate of science and the personality of the scientist come to reflect the common fears, anxieties or anticipations which, because they are shared, become implicit codes determining scientific activity? How much is science enriched or distorted by transcendentalism or mysticism? Bose's life story—and even his brittle self-definition—offers a chance to explore some of these issues.

But before telling the story I must make a caveat. It is, of course, obvious that while the psychosocial pressures to which Bose was subject were fairly typical, his techniques of coping with them were not. Bose by no means offered a packaged solution to the psychological problems of Indian scientists. Yet, however idiosyncratic his responses, he *was* the first distinguished Indian scientist of modern times and the first self-conscious standard bearer of the concept of Indian science. He can hardly be dismissed as a deviation from normal science or as a one-man culture of science. He may not have been an amazing genius, but neither was he a mediocrity who was lionized and blown up into a hero by Indian admirers. His western admirers were, if not more numerous, at least more influential. In fact, some Indian intellectuals sensed early on Bose's diminishing relevance to science and to the institutional growth of science in India, but were afraid of saying so because of his stature outside India. Their reaction was the obverse of that of some others in their country who became enthusiastic about Bose only after his success in the West. Bose was less parochial and less irrelevant to the culture of world science than he may seem today, and he was more typical of Indian responses to western science than Indians may wish to acknowledge. In fact, the phenomenon called Bose can be understood only with reference to certain problems in the culture of science which became central at that point of time.

II

Jagadis Chandra Bose was born in 1858 in East Bengal (now Bangla Desh) in a small, well-to-do Brahmo family.[1] The Boses were originally high status Kayasthas and had owed allegiance to the Shakti cult. Each of these details had its own psychological meaning.

Bose's forefathers came from Bikrampur near Dacca, a place the middle-aged Bose described as a 'producer' of rugged innovative frontiersmen, who crossed seas and scaled mountains 'at their mother's bidding'. Bikrampur 'was not for the weak'; like a spirited mother, she demanded of her children *bikram*—aggressive courage.[2]

[1] On Brahmoism as the first Hindu reform movement of modern times, its beginnings in Bengal in the early nineteenth century, and its attractiveness to urban, high caste, westernized, middle and upper class gentry, see R. C. Majumdar, A. K. Majumdar and D. K. Ghose (eds.), *The History and Culture of the Indian People* (Bombay: Bharatiya Vidya Bhavan: 1963), 10: *British Paramountcy and Indian Renaissance*, Part II, esp. Ch. 13. On some of the psychological sources of the movement, see Ashis Nandy, 'Sati: A Nineteenth-century Tale of Women, Violence and Protest', in *At the Edge of Psychology* (New Delhi: Oxford University Press, 1980), pp. 1–31.

[2] J. C. Bose, *Abyakto*, edited with comments and notes by Pulin Bihari Sen (Calcutta: Acharya Jagadis Chandra Bose Birth Centenary Celebration Committee, 1958), pp. 21–2.

Bikrampur was associated with two other things: first, as Bose was fond of mentioning, it had an ancient reputation as a seat of intellectual activity. It had been a place of Sanskritic and Buddhist learning in earlier times and there were even the ruins of an ancient astronomical observatory to substantiate the faith of the inhabitants of Bikrampur in their long tradition of scholarship. The community had also produced the famous thirteenth-century Buddhist theologian and monk Atish Dipankar, who had carried to Tibet the 'message' of India. This history was often invoked by the inhabitants and by Bose himself to justify their chosen status.

The second association, according to Patrick Geddes, was a certain sensitivity to matters of faith in the region.[3] In the middle ages, Bikrampur had been a centre of Buddhist learning and dissent from the orthodox Sanskritic traditions. With the decline of Buddhism in India, the place became a centre of Hindu revival, although the population of the area all around had meanwhile become Muslim. The result was that the Hindus of Bikrampur in Bose's time felt called upon to maintain, to stretch Geddes' observation, a certain alertness and heightened protectiveness towards their cultural inheritance.

These demands of Bose's birthplace did not coincide with the demands of East Bengal on his self image. If the former gave him the feeling of being chosen, the latter made him feel marked out for second-class cultural citizenship. Before a changed political geography and enforced mass exodus broke through the barriers of local lifestyles and regional identities of Bengal in 1947, the East was known as the backyard of *babu* culture—a damp, marshy, dialect-speaking nest of provincials where ugly *Bangal* ducklings dreamt of becoming elegant Calcuttan swans. In the popular imagination, Bengalis were enterprising, obstinate, and aggressive, but this did not preserve them from the disdain and sarcasm with which everywhere upwardly mobile rurals are regarded by 'polished' urbanites.

The East Bengali response to this stereotype seemed to indicate that the contempt affected them, if only partially.[4] On the one

[3] Patrick Geddes, *The Life and Works of Sir Jagadish C. Bose* (London: Longmans, 1920), p. 2.

[4] Throughout his life, Jagadis Chandra was sensitive about his East Bengali past, particularly his persistent East Bengali accent. See C. C. Bhattacharya, 'Acharyadeb Smarane', *Vasudhara*, 1958, 2, pp. 121–4; also M. Gupta, *Jagadish Chandra Bose, A Biography* (Bombay: Bharatiya Vidya Bhavan, 1964), pp. 6–7. See below an account of his first encounter with westernness and urbanity in Calcutta.

hand, they angrily reacted to such disparagement; on the other, they revealed their deeper identification with their tormentors by ruthlessly rejecting their own authenticity. By becoming second-order *babus* and, with a vengeance, first-order successes, they hoped to gain a greater acceptance of themselves. But if the contempt of others dies hard, self-contempt dies harder; and until political surgery cut off greater Calcutta so completely from the people whose economy and sense of cultural inferiority it had exploited to build its own self esteem, all good Easterners went to Calcutta when they died.

The Boses, despite some lean times, remained solvent and Bose had a comfortable childhood. His hobbies and pastimes were certainly not those of a deprived child. Some later chroniclers, encouraged probably by Bose himself, have tried to give an impression of financial hardship in his early life, but this, on closer examination, proves to be one of those log-cabin-to-White-House myths that recur in the more ambitious segments of a community.[5] It is more reasonable to conclude that, like most other nineteenth-century Bengali stalwarts, Bose had an upper middle-class up-bringing. More real were the fluctuations in the family finances of the Boses, due to his father's adventures in big business which ended in near-disaster. The young Bose *was* exposed to severe economic anxiety, but it was generated more by his father's failure within a competitive, achievement-criteria-dominated, colonial political economy than by actual poverty. There is no way of finding out if it was this which first sensitized Bose to the alternative traditions of apparently stable, self-contained village communities to which he was simultaneously exposed in his childhood. The fact remains that the adult Bose consistently tried to see the attainment of a stable, orderly, simple scheme of subsistence as a specific expression of a wider search for order and simple unifying principles, in nature and in society. The search was reflected in Bose's later conversion to the new versions of Indian nationalism which sang the glories of the traditional economy and decried the horrors of 'crass western materialism'.

The Boses were Brahmos. Monotheistic, anti-idolatrous, caste-denying Brahmoism was still the most creative and intellectually alive Hindu sect. But what had started as a radical movement was

[5] For example, M. Roy and G. Bhattacharya, *Acharya Jagadish Chandra Bose* (Calcutta: Bose Research Institute, 1963), vols. 1 and 2; and Bose, *Abyakto*, pp. 22–3.

by then already showing signs of a defensive rigidity and a tendency to become cocooned within an ideological and moral purism that, within a few decades, was to destroy its ability to withstand the more aggressive Hindu reform movements. Yet Brahmoism could still inspire in its followers a sense of pride—a feeling that they were in the vanguard of social reform, and that they were obliged to 'keep up with' the expectations of some inner, as well as transcendent, authority.

Bose's father was a first-generation Brahmo and his mother never gave up her Hindu orthodoxy. The family lifestyle therefore was still strongly influenced by Shakto high cultural traditions. This may on the one hand have reduced in the family the abrasive qualities which are often associated with membership of a religious protest movement; on the other, it ensured access to some of the deepest and most powerful symbols of Brahminic traditions in Bengal. Add to these the contrast between the rural origins and exposures of the Boses and the highly westernized education and occupation of the father, and you get some flavour of the wide repertoire of cultural themes and role models available to young Jagadis.

The household consisted of Bose's parents, his paternal grandparents, and his five sisters. His brother had died at the age of ten. It appears that the family as a whole was ambivalent towards Bose: together with its adoration of him there was a latent fear that it had given him too much latitude. This could be the ambivalence of the Brahmo subculture itself. Bose was the only son and the youngest in the family, in a society which prescribed preferential treatment for sons because of their economic and ritual role. But Brahmoism built its implicit codes of child training on a rule-of-thumb synthesis of imported Victorian puritanism and indigenous high-caste asceticism.[6] The prudish strictness of Brahmo child-rearing was probably a reaction to the dominant Indian tradition of indulgence and noninterference in the process of early socialization. But not knowing itself to be a reaction, Brahmo socialization rarely freed itself from what it was reacting to, particuarly so in families like Bose's in which some members continued to represent

[6] Towards the end of the eighteenth century and the beginning of the nineteenth, under British suzerainty, this Brahminic asceticism was displaced by an anomic hedonism in greater Calcutta and in the other urban areas of eastern India. Brahmoism can be read as a response to this anomie and as an attempt to reinstate the Brahminic values in a new garb. In this sense, it can be called a particular form of Sanskritization. For a brief discussion of this see Nandy, 'Sati'.

the non-Brahmo traditions within the family. A child often gauged the contradiction; deep down he knew that he had become a battleground between the old and the new, and that he would have to develop a technique of coping with the two irreconcilable but co-existing environments in his universe.

III

What was the actual content of this multiverse of Bose's?

One insight into his early interpersonal world is offered by his favourite *jatra* or folk play. Biographers have already vaguely recognized that this *jatra* draws, as a cherished myth, the outlines of an inner design against which one could examine the realities of Bose's outer world. I shall restate this awareness in my own terms.

The myth was the story of Karna in the Mahabharata. Illegitimate son of the mighty Sun-God, Karna was left to die at birth by a cruel, opportunist, all-too-human mother. He survived to become the proud, aggressive, autonomous, sun-like son of a kindly foster-father—a charioteer who humbly helped him to reach the solar stature and identity that was his by right. Bose admired the parity-seeking 'backyard man' Karna for the mix of good and evil in him, and also because, spurning all temptations, he fought fate with a courage that redeemed even his final defeat.[7] This defeat, the story goes, was ensured when Karna refused to make up with his mother before the climatic fratricidal battle of Kurukshetra. She was then in the enemy camp and trying, he felt, to placate him to save her favourite and legitimate son Arjuna, whom Karna had vowed to fight to the death before he realized his own real origin. Till the end Karna stuck to his humbler identity, in victory as in defeat, and defied the conspiracy of the gods and the original humiliation of maternal rejection by dying as his father's son in battle.[8]

[7] Roy and Bhattacharya, *Acharya*, 14.

[8] Bose's persistent fascination with the personality of Karna was also expressed in his request to Rabindranath Tagore to write about Karna. Tagore's 'Karna–Kunti Sambad' (*Rabindra Rachanabali*, Calcutta: West Bengal Government, 1962, *5*, pp. 578–82) was born in 1899. The work is an imagined dialogue in verse between Karna and his mother Kunti. It depicts Karna's intense sensitivity to the earlier betrayal by his mother, his angry refusal to make up with her even in the face of total defeat, and his mother's earnest attempts to undo the past and build their relationship anew. For two fragments of Bose's correspondence with Tagore on the subject, see his letter of 22 May 1899 in Roy and Bhattacharya, *Acharya*, p. 4, and in Gupta, *Jagadish Chandra*, p. 70.

It was in 1899 that Bose's research showed a new trend. The relation between this and changes in Bose's personality is discussed below (Section VII).

To go by the design of the myth, these cherished—and projected—themes of a grand defeat, pertinacious obstinacy, the conspiracy of fate, personal achievement, and *noblesse oblige* were tied together by the images of a heartless rejecting mother and a fiery male progenitor of all life behind a warm accepting male authority. Psychologically, the son within himself had only the choice of disowning the former and owning up the latter.

We are largely ignorant of Bose's first social relationship. While he and biographers identifying with him have written and spoken at length about his father, they have been reticent—and defensive—about his mother. We only know that her name was Bamasundari and that she remained an orthdox Hindu even after her husband had become a Brahmo. But we have no way of knowing the actual form these differences in faith took within the family.

However, two scraps of information are frequently found in biographies. The first, and the only one to find a place in Bose's own writings, describes Bamasundari's motherliness towards the child Bose and his friends. We are also told by some chroniclers that Bamasundari was severe about Bose's cruelty to animals, and imposed some restrictions on his pastimes and play in childhood. But the episode which biographers most frequently mention is the clash that took place between mother and son when she refused to let the adolescent Bose cross the seas for further studies in England. (Later she relented and even sold her ornaments so that the trip could be made, but probably not before she herself, and the constraints that she represented, had been further associated in her son's mind with traditional Hinduism.)

These scattered references suggest that Bamasundari's indulgent mothering of her only surviving son was mixed with some intervention and restraint. This mix, some informants believe, was partly a result of the stubborn, aggressive ritualism that pervaded Bamasundari's entire lifestyle. The belief is endorsed by the sensitive observations of some members of the Bose family; they help us to locate, with tolerable reliability, those crucial undercurrents of mothering which formal biographies merely hint at.[9] The episode of the sea-trip to England, they would have us believe, was typical of Bamasundari's 'peculiar' ability to antagonize her son,

[9] Bose also had something to say about his mother: his studied silence about her was accompanied by repeated and almost obsessive references to protective and aggressive motherliness and to nurture and succour in his work—sometimes explicit, always implicit.

an ability that often elicited from the latter an angry, fiery reaction or a form of free-floating aggression that was directed not towards her but towards the world at large. It was not so much Bamasundari's 'old-fashioned ritual self', as her son once put it,[10] as the edge given to it by her crusty obstinacy that he found so exasperating.

A nephew of Bose, a distinguished scientist who was close to his uncle, also remembers Bamasundari as 'definitely neurotic', a lay diagnosis referring to the pronounced symptoms of obsession-compulsion that she reportedly showed. This, of course, was nothing uncommon in Bengal and, for that matter, India. Euphemized as *shuchibai*, or compulsive ritual cleanliness of body and mind, the symptoms were so 'popular' that they perhaps could be called a culturally sanctioned posture of certainty, to face deeper uncertainties generated by common anxieties. By turning symptoms into tolerable angularities, the society evidently institutionalized certain behaviour patterns that in some other societies would have remained a part of psychiatric symptomatology. The compulsiveness of Bamasundari, if it can be called this, could have been an accentuated form of some aspects of Bengali normality; and her kind might have been, if not quite the rule, scarcely the exception. What must have aggravated the stress was the flamboyant bouts of anger that accompanied her finicky demands. Some have linked the mother's outbursts of rage at every failure to meet her criteria of ritual performance and purity, to the perfectionist, equally finicky, scientist-son, throwing tantrums at his own and others' inability to meet his standards of excellence.

At the same time, his mother remained for Bose a symbol of indomitable will, abrasive defiance of external stimuli, succour, and fixedness of purpose.

> Man is not a servant of fate, within him there is the power by which he can become independent of the external world.... This is the means through which he will ultimately triumph over physical and mental weaknesses.... Inner power is self will!... At which stage of life is this power born?... At which stage of life does this power to fight grow?... I was thrown small and helpless into this sea of power at birth. Then outer power entered inside me to nurture my body and to help it grow. With mother's milk, affection, pity and sympathy entered my heart and the love of friends made my life flower. Power has collected inside me in response to bad days and

[10] Bose, *Abyakto*, p. 125.

external aggression and with this power I have fought against the external forces.... Life takes form due to the power struggle between the inner and the outer. At the source of both the outer and the inner lives is the same *Mahashakti* who powers the nonliving and the living, and the atom and the universe.[11]

As we know, *Mahashakti*, the great power, is also the ultimate maternal principle in Indian, particularly Bengali cosmology; it invokes mother-deities who combine traditionally the ultimate in benevolence with the ultimate in terror.[12]

It was this image of a mother, at once benevolent and terrorizing, that found expression in young Bose's favourite myths and folklore. Long afterwards it found even clearer expression in his public idiom, in his closeness to reformers and religious leaders who invoked this image as their major symbol, in his return to aspects of the Bengali projective systems presided over by a series of powerful mother-deities, and in the distinctive culture of science he tried to build around him.

We do not know how successful Bose was in handling his ambivalence towards his mother. One defence was perhaps to discover a continuity between his mother and the family tradition of mothering. The discovery might have been made possible by the presence in the household of an authoritarian grandmother, caring and restrictive at the same time.[13] Even more useful might have been the history of a quasi-mythical great-grandmother. As Bose tells the story, she was widowed at an early age and came with her young son to stay with her brother. Though an affectionate person, she was a stubborn fighter against the difficult times she was facing. One day her young son, terrorized by his teacher, came to her for protection. According to Bose, 'the affectionate mother immediately turned aggressive, tied the hands and feet of the son, and handed him over to his teacher.'[14] These imageries of a nurturing but violent mother, of a child seeking a mother's protection that is denied, transient male authorities who terrorize,

[11] Ibid., pp. 191–4.

[12] See on this subject Ashis Nandy, 'Woman Versus Womanliness in India: An Essay in Cultural and Political Psychology', in Nandy, *At the Edge of Psychology*, pp. 32–46; pp. 301–15; also Nandy, 'Sati'.

[13] 'Jagadis Chandrer Jivani', *Vasudhara*, 1958, 2, pp. 124–35, 126. Reportedly Bose's grandfather was a more accepting and indulgent figure than his grandmother. It seems that the grandparents, too, partly replicated the parents for Bose.

[14] Bose, *Abyakto*, p. 121.

and a father who cannot provide protection because he is dead or absent were to recur in Bose's later life. At the end, this imagery of a widowed mother, looking forward to her son's success, was to blend with Bose's concept of a motherland dependent upon his own success in the cruelly competitive world of science.

All this means that differentiation in the image of the mother could become, in Bose's case, the matter of a more permanent split. Mother love was to denote for him both 'nurturing kindliness' and 'annihilating power' for all time to come.[15] When used as parts of an idiom, this was to have special appeal in a culture dominated by a maternal deity who, in her multiple incarnations, represented the extremes of benevolent benignity and incorporating terror. Bose's later drift towards the world-view of pantheistic Hindu orthodoxy, which horrified many pietistic Brahmos, was, therefore, more than a considered, rational choice; it bridged the loves and hates of his infancy and the preferences of his adulthood.

Bose's mythical ego ideal Karna would have perished but for the intervention of a kindly foster-father. And the scientist's later childhood and youth were spent in a tenacious attempt to use his apparently conflict-free identification with his father as the core of his self, perhaps to disown his identification with his mother and her ritualistic, magical, domineering self.

Bose's father Bhagwanchandra was considered a cultivated man who, in Bose's younger years, was employed mainly in small towns in rural areas. To be cultivated in nineteenth-century urban Bengal had a significance which it is difficult to convey today. The idea of a general education no longer leads to aspirations of the same kind on the part of middle-class Indian parents, and professionalism and specialization have deprived work and knowledge of their earlier connotations of being conducive to the production of a rounded personality. Even Bengalis who, like the Victorians, once made a fetish of a general education, are no longer fascinated by the concept of a gentleman who by his ambidextrous qualities validates his gentility. Nothing denotes the older concept of gentility better than the fact that the elder Bose was, in addition to being a successful administrator, an entrepreneur, an amateur physicist and biologist, a promoter of technical education and part-time engineer, a devout Brahmo with an almost professional interest in the theology of Vedanta, a sportsman, and a part-time social worker.

[15] Ibid., pp. 140–1.

He also nurtured ambitions of becoming a writer and a nationalist leader. It is evident from the admiration that this wide range of roles and interests evoked in Bhagwanchandra's contemporaries that the ideal Bengali gentleman was still a person whose multifarious talents idealized—and thus negated the anxiety of—the role confusion typical of those changing times. We shall see how this confusion, and the anxiety associated with it, later forced his son to look beyond the identification represented by the father towards the clearer self-definition offered by the mother.

Before turning to the content of Bose's relationship with his father, let me summarize what little is known about the earliest interests and pastimes of young Jagadis. These are important because they provide a clue to what he later made out of his relationship with his father. The child Bose's curiosity and enthusiasm were, we are told, unbounded. All types of growing, self-propelled, and living objects fascinated him. Together with this interest in life and life processes was also a preoccupation with violence. He played with models of battleships, cannons, and cannon shells, and trapped and killed birds (his father was permissive in this respect, his mother strict). The beginning of his lifelong interest in aggressive wild life can also be seen in his childhood practice of catching snakes and playing with them. Some of this concern with symbolic and not-so-symbolic violence was, even at that stage, linked to his creative efforts. For example, he made, with the help of unskilled labourers, the brass model of a cannon which was considered to be a remarkable feat of engineering.[16] One suspects that Bose was already using his aggressive fantasies to acquire mastery over his environment, by exploration, and experimentation.[17]

As he grew up he took to the 'manly' sports of shooting, riding, and *shikar*. The *shikar* was mainly big-game hunting and later on he would remember his tiger-hunts with some nostalgia.[18] Perhaps

[16] This model was the first intimation of Bose's superb skill at scientific instrumentation. It also provided another link between the amateur engineering of Bhagwanchandra and the applied physics and experimental physiology of his son.

[17] Phyllis Greenacre, 'The Childhood of the Artist: Libidinal Phase Development and Giftedness', *The Psychoanalytic Study of the Child* (New York: International Universities, 1957), *12*, pp. 47–52.

[18] Bose, *Abyakto*, pp. 185–6; and J. C. Bose, 'Dedicated Life in Quest of Truth', in P. K. Chatterji (ed.), *The Presidency College Magazine* (Calcutta: Presidency College, 1964), *Golden Jubilee Volume*, pp. 64–73. See also Roy and Bhattacharya, *Acharya*, pp. 18–19. Throughout his life Bose continued to decorate his drawing room with stuffed animals he had shot in his younger days. These symbols of aggression against nature

hunts were the occasion for him to externalize some of his deeper violence—by pursuing and killing wild aggressiveness, he could destroy some of those inner furies which sought an outlet in his games, pastimes, and creative products. The fact remains that the adult Bose, even when he had turned a militant pacifist, continued to feel proud of his record as a big-game hunter.

Bose's companions in these ventures were often adults and this frequently brought out his fears of being underrated and his early search for parity. A typical incident was his participation in a riding competition; though he was far behind throughout, he refused to give up and was injured badly in the process. Two similar incidents, involving his obstinate search for recognition and being successful in a competition, were a brawl with a gang of Anglo-Indian classmates and a boxing contest he won against heavy odds. Both took place in his teens in Calcutta. In each case his father openly admired Bose's exploits and, on at least one occasion, publicly congratulated him on his tenacity. One suspects that the westernization of Bhagwanchandra went beyond his 'civilized' interests, hobbies, and occupation. It had ramifications in some acceptance of aggressive competition and competitive achievement.

Bhagwanchandra also took a special interest in his son's education. It was he who introduced him to botany and physics. This direct intervention in the son's education was not common in those times in a subculture which promoted some degree of distant paternalism. But it paid dividends. The intimacy and mutual respect which grew between the father and son was to act as a fulcrum of the son's personality for a long time. Of course, there was also the relatively rare circumstance of a modernized family moving from place to place in rural and semi-rural Bengal, according to the vagaries of a transferable government job. The father may have been partly forced to intervene in the son's upbringing. But there was nothing forced about the warmth and permissive friendliness with which Bhagwanchandra handled this duty. The consequence was a deep sense of personal obligation in the son, and also a developed sense of *noblesse oblige*.

The best-known example of this has already become a popular myth among the Bengali élite. The story goes that Bhagwanchandra at one time pioneered a number of industrial ventures, which were

---

were to become rather anachronistic in his later life. See below.

to yield fabulous profits later but to him were merely a source of acute financial embarrassment.[19] This landed him deeply in debt. From this distance in history, we have no means of knowing where the decision-making went wrong; we have only the son's expressed belief that it was his father's risk-taking adventurous defiance of his safe vocation and secure income that was being punished by fate. It is evidence of Bose's idealization of his father that when he grew up he took enormous trouble to repay loans which even the creditors had written off as bad debts. Not merely that; he diligently publicized his justification for his father's business failures. In fact, his homage to Bhagwanchandra's entrepreneurship and achievement-frustrations remains one of his most maudlin public speeches.[20] He notes therein the similarity between the 'magnificent defeats' of his father and those of Karna, his favourite mythical champion of lost causes,[21] but naturally fails to see the analogous role he selected for himself, and often played, with such dramatic fervour.

Does the analogy with Karna mean that Bose unconsciously perceived his father to be, like him, a victim of a less-than-benevolent maternal authority? Was this perception endorsed by the self-willed grandmother staying with the family, or was it merely a projection of his own relationship with Bamasundari? Did this perceived victimhood seem to him to have a sacrificial content? We have no means of knowing. We only know that the image of his father as a co-victim of maternal aggression persisted. 'He was born before his time', Bose stated in his speech. 'It is on the ruins of many lives like his that the greater India of the future will be founded. I do not know why it should be so; all I know is that Mother Earth is hungering for such sacrifices.'[22]

Bhagwanchandra was an active Brahmo and Bose discovered early in life that his father's religious enthusiasm was associated with a lively interest in science and technology, an amalgam of interests that was to become a formal synthesis in the son much later. (This synthesis was sanctified in his—a 'monotheistic'

[19] According to one version, these industrial ventures involved tea and jute. It must have taken some remarkably persistent inefficiency to fail in a business involving these two goods, which within a decade or two became the two most important industries in India. Bhagwanchandra, some may conclude, was a dyed-in-the-wool Bengali gentleman.

[20] Bose, *Abyakto*, pp. 121–34.

[21] Roy and Bhattacharya, *Acharya*, p. 8.

[22] J. C. Bose, quoted in S. P. Basu, *Nivedita Lokamata* (Calcutta: Ananda, 1968), *1*, p. 569.

Brahmo's—incongruous, life-long devotion to Vishvakarma, the second-order Hindu god of technology and scientific creativity. In particular, Bhagwanchandra was well known as a builder of instruments. This is an area in which his son too excelled, and later even the most violent critics of Jagadis Chandra's scientific theories admitted his skill in scientific instrumentation.

It was his father, again, who handed the child Bose over to two elderly servants for training, protection, and upbringing.[23] One of them, who became very friendly with Bose, had been a robber in his younger and more glorious days, and was once condemned to imprisonment by Bhagwanchandra himself. He often regaled his charge with his 'tales of bravery' and Bose was truly impressed when this retired bandit, using half-forgotten skills, saved the entire Bose family from pirates. Bose never forgot his bandit friend, and the theme of control of latent aggression, by aggressively protecting an erstwhile target of attack, was to become an important one in his life and work.

The other servant who looked after Bose was a retired army sepoy who taught Bose how to use a rifle, which, as we have seen, later came handy in his safaris.

Bose's nationalist father sent him for his early education to a *pathshala* or village school which he had himself founded. It was still possible in Bengal to combine government service with ardent nationalism and Bhagwanchandra, one may say, was one of the last to make a success of this combination. Nevertheless, his decision to send Jagadis to an indigenous school shocked many. Firstly, nobody expected him, a *babu*, to take seriously a school which he had established as a social service. Secondly, in those days children of highly-placed civil servants were invariably sent to English medium schools.

But Bhagwanchandra's decisions, when they did not relate to

[23] The role of servants in the socialization of nineteenth-century *babus* can be the subject of an interesting study in transitional institution-like structures. Servants represented the Bengali élite culture's new and anxious recognition that children required more supervision, control, and intimacy with adults than was provided for in the older model of socialization. By meeting some of these needs, and by validating and invalidating earlier interpersonal experiences of the children, these servants significantly influenced the transition from infancy to adulthood. And to the extent that they symbolized the parents or mediated parent–child relations, they played a vital role in the child-rearing system of the subculture. For a vivid account of the crucial role of servants in child-rearing in a modernist Bengali élite family, see Rabindranath Tagore, *Chhelebela* (Calcutta: Viswabharati, 1944).

industry and commerce, bore the stamp of pragmatism and vision. The exposure to the progeny of humble village folk and to their lifestyle that young Bose had in school was crucial to his developing self. First, his peers in school were his social inferiors and they never challenged the sense of being chosen that Bose had as an only son, in a culture which emphasized the importance of sons. On the contrary, these peers might have negated the threat to his self-esteem that Bamasundari's control represented. Second, Bose's schooling endorsed his sneaking respect for the stability and orderliness of the traditional systems. At least Bose later claimed this.[24] Perhaps it also helped him internalize the concept of a reference group to which he would at least apparently veer round, whenever threatened by criticism or neglected. Bose himself was clear about why his father had sent him to an indigenous school:

> Now I know why my father put me into a Bengali school. There I had to first learn my mother tongue, think in that language, and I got acquainted with the national culture through a national language; I also learnt to think of myself as one of the masses, and no feeling of superiority separated me from others.[25]

Despite such exposure, Bose remained a lonely child. This loneliness, his poor knowledge of English, his rustic East Bengali mannerisms, and the sudden fall from the status of being a magistrate's son to that of being an Indian amongst Anglo-Indian students and teachers, created a rather uneasy situation for him when he entered St Xavier's School at Calcutta in 1870. We have already mentioned the street brawl and the boxing contest; these incidents were the overt expression of an environment that further sensitized him to his nationality and colour, and helped him to associate his personal humiliations with the political status of his people.

Things improved when he went to St Xavier's College on a scholarship. There he developed a decisive, long-lasting intimacy with a Jesuit priest, Father Eugène Lafont (1837–1908). Lafont was the best-known professor of experimental physics of his generation in India and was one of the founders of the Science Association of Calcutta. He is still remembered as the 'father of

[24] Bose, *Abyakto*, p. 125.
[25] Ibid., pp. 125–6.

science in Bengal'. A Belgian trained in philosophy and the natural sciences, he had arrived in Calcutta towards the end of 1865. When Bose joined St Xavier's College in 1875, Lafont was its Rector and already the Father was regarded as a pioneer in science education in India. His public exhibitions and lecture demonstrations on science were highly popular. The popularity was due not only to Lafont's personality; science was now in the air. Such was the curiosity of the *babus* of Calcutta about the new science that a contemporary report on Lafont's lecture series says:

> Notwithstanding rain and lightning, a pretty fair attendance of native gentlemen gathered around the lecture table to hear the exposition of Dalton's atomic theory and witness some experiments, illustrating the general principles of matter.... Throughout the series, the gentlemen were most assiduous and punctual in their attendance.[26]

Jagadis Chandra was inspired by Lafont's presence and this new enthusiasm for science. At first he was interested in 'natural history' and wanted to become a botanist. Lafont encouraged him to turn to physics. Bose studied physics, but combined it with Latin and Sanskrit, perhaps impressed by the mix of classics and sciences in Lafont's life. It was not only in the choice of discipline that Bose was influenced by the Jesuit. Lafont's lecture series might even have helped Bose internalize an atypical concept of a scientific audience. The series could have made him aware of the possibilities of a slightly exhibitionistic public exposition of science in a society with a strong oral tradition. Here, therefore, was another fatherly figure who combined physics with religion and metaphysics. And for Bose the experience was another stage in an unfolding intellectual identity stretching from infancy to adulthood.

Bose graduated in 1879. Like other young men of his generation and background, he started toying with the idea of becoming a civil servant. This was a highly valued occupation which, in those times, had a special attraction for the ambitious contemporaries of Bose. Also, members of the Indian Civil Service were paid well and Bose wanted to lessen his father's economic burdens. But the magistrate-father vetoed this and encouraged his son to become an academic. Instinctively, Bhagwanchandra pushed his son towards a self modelled on his—Bhagwanchandra's—idealized but partly fulfilled

[26] A. K. Biswas, 'Rev. Father Lafont of St. Xavier's College', in *Science in India* (Calcutta: Firma K. L. Mukhopadhyay, 1969), pp. 67–84.

self-definition as an intellectual, rather than on his more superficial commitment to a government job.

Bose's early role confusion, a paternal inheritance, pursued him even to England where he lived from 1880 to 1884. During 1880–1 he studied medicine but gave it up, partly because of ill health. In 1881 he joined Christ's College, Cambridge, on a scholarship to study botany, and graduated from both Cambridge and London Universities in the natural sciences. Throughout his undergraduate days, he took a variety of courses without specializing in any. However, his relations with the professors of physics and botany were decidedly deeper. In particular, his professor of physics, Lord Rayleigh, served as another Father Lafont.[27]

One does not have to be a psychologist to conclude that Bose was trying to build his self-definition, even at that stage, on his relatively conflict-free identification with his father. And, in offering him confusing and often contradictory sublimations and ego defences, that identification reveals what we have already recognized: a diffusion of identifications in the father himself. This confusion, that had dogged his father's steps all along, now picked up the son's scent too, turning careers into pastimes and pastimes into careers, as in the earlier generation.

IV

These were the assets and disadvantages with which Bose came back from England in 1884 to become an acting professor of physics at Presidency College, Calcutta. In getting the appointment, he was helped by Dr Fawcet, a benevolent elderly English educationist who was working as the Postmaster-General, and by Lord Ripon, the Viceroy of India and one of those historical figures who by their sheer personal goodness often help consolidate the oppressive systems of which they happen to be apparently unwilling parts. Both, by taking a personal interest in such a triviality, probably reconfirmed Bose's inner image of essentially benevolent public authorities.

The appointment was made despite the protests of Alfred Croft,

[27] A. Home (ed.), *Acharya Jagadis Chandra Bose* (Calcutta: Acharya Jagadis Chandra Bose Birth Centenary Committee, 1958), p. 6. Note that for the second time in his life, Bose had moved away from a subject akin to 'natural history', his first love, to physics because of his special relationship with a fatherly figure more interested in the physical sciences. He was to come back to 'natural history' again.

the Director of Public Instruction in Bengal, and C. H. Tawney, the Principal of Presidency College. Both supported the strongly-held belief of the Education Department of the Government of India, that Indians could be excellent at metaphysics and languages but not in the exact sciences. This belief was often a matter of genuine conviction, something that was ignored by the nationalists who wanted to make a *cause célèbre* out of Bose's difficulties. Thus Tawney, who so openly protested against Bose's appointment, was also a great admirer of Indian achievements in metaphysical and religious thinking and a devotee of Shri Ramakrishna, the great mystic and religious preceptor.[28]

Tawney's position was based on two typically European assumptions, both by now endorsed by sizeable sections of the Indian élite. First, there was a wide gap between the scientific achievement of the West and that of India fixated, as it appeared to many, at the level of medieval technology. Second, the natural sciences were more important than other disciplines, and in the more crucial disciplines Indians were not as good as Englishmen. The Crofts and the Tawneys looked down upon Indians only in this indirect sense. It was India's westernized middle classes which, having internalized the western concept of the primacy of science, felt humiliated because they felt they were inferior in the natural sciences and because they no longer believed in their own culture's hierarchy of knowledge.

As interesting as the process of appointment was Bose's place of work. Established in 1817 as the Hindu College, Presidency College was the first Indian college to impart western education. In Bose's time, it was by common consent India's best-known college, the most prestigious training ground for the Bengali *babus*, and their most famous gateway to the Occidental currents of thought. It was in this institution that the progeny of the Bengali aristocracy earned their spurs as modern citizens of the world exactly as their progeny, today, learn the very latest in imported theories of progress.

Unfortunately, the college at that time had practically no facilities for empirical research in science, but fortunately, this was a challenge Bose enjoyed taking up.[29] He began to devise his own instruments and laboratory aids. Bose also probably enjoyed the

[28] Basu, *Nivedita*, vol. 1, p. 571.
[29] Ibid., p. 8.

protracted quarrel with the government Education Department in which he got entangled. The quarrel was about differences in the salaries of British and Indian teachers. The obstinate East Bengali, sensitive to even the smallest slight, protested by foregoing his pay. Though he finally won when the disparity was removed after three years of unremitting struggle, his family life and personal work suffered another round of instability, brought about by the vagaries of the colonial educational system.

Bose fought his battle in a predictable fashion. Faced with a hostile authority that did not conform to his concept of what a paternal authority should do, he coped with his problem exactly as one would have expected him to: he appealed to the higher authorities, trying to arouse their sense of rationality, justice, and benevolence to grant him the parity that he had earned through hard work and skill. His success in this instance was probably the final validation that ensured the more or less cordial relations between him and the British authorities throughout his life, cemented later by that well-known token of paternal reward in British India, a knighthood.

However, the scars of the conflict may have remained, to make him rather more suspicious of western systems of education and to revive his awareness of his own deep-seated concepts of learning and vocation. And the search for personal parity that had engaged him for years, gradually began to take the shape of a search for national parity in this changed context.[30]

But Bose enjoyed his work all the same. This was his first experience of being a teacher and wielding authority. By all accounts, he was a success, friendly and egalitarian with his students.[31] Both the friendliness and egalitarianism, however, were confined to this period of his life and did not extend beyond his professional prehistory.

Till this time, Bose had not achieved any academic distinction, though some of his teachers agreed he had promise.[32] Even as a professor, he was concerned with hobbies such as photography and the building of simple laboratory instruments for his college,

[30] That he did not widen and politicize this particular version of nationalism, as he could surely have done afterwards as a national hero, only goes to show how much he preferred to, and could, keep his relationship with the authorities conflict-free.

[31] Roy and Bhattacharya, *Acharya*, p. 29; this is based on Ramananda Chatterji's description in 'Kashtipathar', *Prabashi*, November–December 1932 (Agrahayan, 1939).

[32] Ibid., p. 18.

rather than with sustained serious research. The waywardness and unfocussed passions of his youth, to which some of his biographers refer, also found expression in the aimlessness of his early professional life.

In 1887 Jagadis Chandra married; it was an arranged marriage to a woman from an orthodox and illustrious Brahmo family with a pronounced reformist and religious bent. It is said that his wife, Abala, was a particularly devout Brahmo whose insistence on regular prayers and meditation revived forgotten memories in Bamasundari's son.

Abala had studied medicine for four years before her marriage. Thanks to this exposure, she was not only able to share Bose's interest in his scientific hobbies, but also to encourage him to take up serious scholarship. Finally, when he actually did so—I shall describe the episode below—she took an active interest in his social and intellectual needs. Much light is thrown on Bose's adult personality by the nature of his conjugal relationship and the role which Abala, by all accounts, played in his life. Geddes, who knew them first hand, says:

> ...Hers has been no simple housewife's life,...not only appreciating her husband's many scientific problems and tasks, and hospitality to his students and friends, but sharing all his cares and difficulties.... For his impassioned temperament—in younger days doubtless fiery, and still excitable enough—her strong serenity and persistently cheerful courage have been an invaluable and ever active support....[33]

Bose's nephew, D. M. Bose, also speaks of Abala's calm strength and unruffled temper, and of her tact which compensated for her husband's poor interpersonal skills when dealing with important scholars.[34] Her serenity in a charged situation, Geddes says, lay in her ability to accept and modulate Bose's basic combativeness—'like the fly-wheel steadily maintaining and regulating the throbbing energies of the steam-engine'.[35] This firmness and control came overlaid with conspicuous submissiveness.

The combination was important for someone with Bose's 'pre-history': particularly the distance and ambivalence that characterized his relationship with his mother. By being motherly towards

[33] Geddes, *Life and Works*, pp. 218–19.

[34] D. M. Bose, 'Abala Bose, Her Life and Times', *Modern Review*, June 1966, pp. 441–56, esp. pp. 445–54.

[35] Geddes, *Life and Works*, p. 219. Also Roy and Bhattacharya, *Acharya*, p. 18.

Jagadis, his relatives and friends, by part seriously and part playfully assuming the role of a firm, moral authority, Abala invoked the memories of a relationship that had found its bearings in the socializing fantasies of his culture—in the rich, early tradition of mother-image-dominated Bengali myths. Here was a non-threatening, apparently controllable mother, calm, supportive and doting, and yet more convincing by virtue of being firm, independent and self-sufficient.[36] The two faces of Bose's inner mother at last showed signs of becoming one in this sensitive, humane, determined woman.

There was support for this relationship from three other sources. In a society where childlessness provoked personal and collective anxieties about the reproductive capacities of nature and the survival of the family, 'barrenness' often forced a couple to close ranks. For together they had to face a society that stigmatized them, particularly the wife, as a symbol of inauspiciousness and marital failure. Brahmos were more liberal in this respect than orthodox Hindus, but there must have been only a partial protection against the messages coming from the larger society. Second, after a decade of marriage, from 1897 onwards, continuous ill-health in her brother's family began to require Abala's presence and intervention. At first she bore partial responsibility; later she took charge almost fully. So much so that the children of the family came to look upon her as a second mother. Whether or not this long contact with an ailing family deepened her 'maternal instincts', as D. M. Bose believes,[37] it confirmed an existing pattern. Thirdly, Abala's sensitivity to, and empathy with Bose's personal insecurities increased their mutual dependence. There is a well-known anecdote that illustrates this. Bose, alwalys seeking nurture, used to ask children who visited him to declare how much love they had for him. Not till they had spread their hands out fully to indicate their total love for him would he be satisfied. Abala knew this and

[36] The maternal role played by Abala is described by almost every biographer of Bose. For a brief account see Gupta, *Bose*, pp. 85–7. As we shall see below, the pattern was confirmed by the motherly women Bose gathered around himself, particularly Ole Bull and Nivedita, two Irish women who had opted for the Indian way of life. (Geddes, *Life and Works*, Ch. 17, esp. p. 221). But it was not blind motherly love that Bose sought. Both these women as well as Abala combined nurture with power, control, understanding and acceptance. However, Abala, by adding to this combination a certain intrusive firmness, completed the image of a phallic mother. Her imposing, somewhat hairy, heavy build might have contributed to this image.

[37] D. M. Bose, 'Abala Bose', p. 445.

would encourage some of her grand-children to visit Bose by rewarding them with cookies.

The death of their only child in infancy in 1902 made Bose even more the locus of Abala's attempts to cope with her 'frustrated maternity' and made the two of them 'more completely...one'.[38] Symbolic maternity now became more critical for her, and her husband's deep need for succour became more than ever a vital necessity. In this environment, Bose's infantile fantasy of being able to appease and get love from a maternal source by compulsive orderliness and purity and by searching for an all-embracing order attained increasing importance. The gaining and holding of love through finicky neatness and order in scientific thought, imagination, and in day-to-day behaviour (for example in dress, food, work-routine, and housekeeping) now became a matter of internal order and even compulsion. The bridge between the disorganized, defiant, angry, wayward youth and the proponent of a unifying theory of life had at last been built. 'The present', Bose had said in 1894 in the first sentence of his first publication, was 'made of the past', and he often found them 'separated by an inescapable barrier'.[39] A few sentences later he corrected himself; in the fresco of a mother nurturing her baby he discovered 'the bridge between the past and the present, built on nurture and motherly love'.[40]

Absolute order, however, can be costly for a thinker, for it is only a short step from order to ritualization. The link between orderliness and compulsion in rituals, by inviting attention to the uncertainties underlying the latter, only warns one against being seduced by the pseudo-certainty of the former. One can thus trace one's steps backwards, and examine each ritual as a specific attempt to impose order and claim certainty where the forces of disorganization and uncertainty are at their most powerful and harassing. The elements of the magnificently rigid certainty that we find in the later Bose were therefore also a way of coping with a long sequence of uncertainty and deep fears of uncertainty.

V

Bose's father died in 1892, his mother in 1894. We have no record of how the son reacted to these deaths. Given the nature of the

[38] Geddes, *Life and Works*.
[39] Bose, *Abyakto*, p. 1.
[40] Ibid., p. 3.

mother–son relationship, one suspects that it was the second death that aroused intense feelings of guilt and fear of his own destructive wishes. In any case, a great change occurred in Jagadis Chandra at this time, and he started his researches in the same year. The turning point was his birthday, 30 November 1894. That day he vowed, persuaded by Abala, to take up scientific research seriously—to further knowledge by unravelling the mysteries of nature, *prakriti*, as both nature and the feminine principle are called in Bengali and Sanskrit. This marked, his nephew thinks, a more or less permanent truce with the conflicts 'which had been going on in the subconscious region of his mind.'[41]

Thus, with the help of his wife, Bose the wayward child and confused adolescent was reborn (it was his birthday) into a new life where work identification and professional selfhood could be built on a rearrangement of earlier identifications. The fight against the temptation to see science as a hobby—his father's hobby—was joined with the newly-found weapon of single-minded absorption in his scientific career. But not before a new, intimate motherliness—his wife's—had helped him partly to accept his rejected identification with his mother. He could now afford to search for evidences of 'motherliness in the steel frame of inexorable orderliness' in a ruthless *prakriti*.[42] In 1894 he wrote: 'Everybody cannot see mother's love in the heart of nature. What we perceive is only the projection of our minds. The things on which our eyes rest are merely the pretexts.'[43]

Some semblance of this search for a benevolent mother in the heart of nature continued for another five years, till another form of motherly nurture helped him to break into a new kind of research into *prakriti*. But the first breakthrough towards serious orderliness—his mother's orderliness—had been made. Though he walked for a while what he later called the safe path of research in electrical waves, he had located his personal projective medium in scientific research—a medium that would be true to him in response to his own fidelity to his deepest self.

Bose looked upon the first five years of his research as a preparatory stage—a prehistory marked by the discovery of instruments, rather than by daring theorizations. This view agrees

[41] D. M. Bose, 'Abala Bose', p. 446; Home, *Acharya Jagadis*, p. 7; Gupta, *Bose*, p. 23; Roy and Bhattacharya, *Acharya*, pp. 32–3.

[42] Bose, *Abyakto*, pp. 3, 133.

[43] Ibid., p. 4.

with the opinion of D. M. Bose and Amal Home, who have divided Bose's research career into roughly three phases:[44] 1894–1899: Production of the shortest possible electromagnetic waves (up to 5 mm) and verification of their quasi-optical properties; 1899–1902: Study of some similarity in responses in the living and the non-living; 1903–1933: Study of response phenomena in plants, the complexity of whose responses lies intermediate between those of inorganic matter and of animals.

I am not competent to evaluate Bose as a scientist, and it is difficult to get a comparative assessment of the three periods from scientists who are no longer sufficiently interested in Bose to devote themselves to all his work with equal seriousness. D. M. Bose, a rare exception, believes that the first phase has the surest scientific relevance to present-day physics and biology, followed by the third and then the second. Others make more or less similar assessments.[45] Bose himself, however, considered the second phase of work, from 1899 to 1902, the most important. After completing his *Responses in the Living and the Nonliving*, he wrote: 'My task is more or less complete; in future I shall merely have to sit idly in an atmosphere of uncertainty.'[46]

Thus while Bose himself and his Indian and western admirers valued the second, third, and first phases in that order, some of his contemporary western scientists' evaluations were not substantially different from the present-day assessments of both Indian and western scholars. The virulence of Bose's detractors notwithstanding, his belief in the anti-Indian parochialism and conspiracy of western scientists might have been based both on a realistic perception of outer hostility and on his and his society's mood. Perhaps Bose and his disciples sought, in the perceived hostility of the West, another justification of their own growing hostility to the West and of their own science.

The researches of Bose's first phase were well received. In 1896,

[44] D. M. Bose, 'Jagadish Chandra Bose (1858–1937)', *Transactions of Bose Research Institute*, 1958, *22*, v–xv; also Home, *Acharya Jagadis*, Part I.

[45] Gopal Chandra Bhattacharya, a botanist who worked under Bose for seventeen years, has a different periodization to offer. He believes that the second phase, culminating in the publication of *Responses in the Living and the Nonliving* in 1902, was the most creative period in Bose's life. Bhattacharya divides Bose's work into two stages. All the significant work according to his scheme was published before 1906; all the trivia afterwards. As Bhattacharya's periodization cuts across the others', his assessment of Bose as a scientist is not strictly comparable with that of anyone else.

[46] Roy and Bhattacharya, *Acharya*, p. 95.

the University of London awarded him a D.Sc. for his contribution to physics and the Government of India sent him to England on a lecture tour, the first Indian scientist to be so deputed. Bose did well in England. His speeches on electrical microwaves at the advanced British centres of learning were applauded by some of the best-known scientists of the time, including William Kelvin, Joseph Lister, and William Ramsay. Some of the more sedate dailies and periodicals, like *The Times* and *The Spectator*, also found his work strikingly original, even sensational.[47] The English tour was followed by lecture-demonstrations given to a number of learned societies and universities on the Continent, which were also highly praised. Bose apparently was well set to become a celebrated physicist, if not an outstandingly creative one.

But, he knew, 'success could be cheap and failure great', and his—a Brahmo's—'goddess lifted her children to her bosom only when they returned to her defeated.'[48] Failures therefore were cushioned:

> What are you afraid of? That you will not reach your goal even at the cost of your entire life? Do you not have the slightest of courage? Even a gambler stakes his life's earning on a throw of dice. Can you not stake your life for a grander game? Either you win or you lose![49]

Thus in 1899, the vision of a grand defeat and what could be the paradoxical fear of negative success,[50] made Bose cross the conventional western boundaries of scientific disciplines to enter the field of botany, a subject he had formally studied only in his undergraduate days.

A number of explanations for this shift can be suggested. One is that Bose's work on short electrical waves had till then mainly taxed his skills as an applied physicist. And instrumentation was, as we know, Bose's forte. But his researches were now leading him towards more complex mathematical work for which he had no aptitude. His students say that he dreaded mathematical details, and he himself later ascribed the change-over—great

[47] See details in Home, *Acharya Jagadis*, *passim*; and Roy and Bhattacharya, *Acharya*, Ch. 11–15. Also Geddes, *Life and Works*.

[48] Bose, ***Abyakto***, pp. 132–58.

[49] Ibid., p. 131.

[50] The concept of negative success is Erik Erikson's. See his *Young Man Luther* (New York: Norton, 1958), p. 44.

rationalizer that he was—to the over-mathematization of modern physics.

However, there were less mundane reasons, too. Bose had now come to believe that 'plant life was merely the shadow of human life.'[51] And he was looking for a projective medium where the objects of his inquiry would be more recognizably the living, feeling victims of the environment who needed humane intervention and protection:

> I once did not know that these trees have a life like ours, that they eat and grow. Now I can see that they also face poverty, sorrows and sufferings. This poverty may also induce them to steal and rob.... But they also help each other, develop friendships...sacrifice their lives for the sake of their children.[52]

His heightened concern with violence and its control also contributed to the change. Shankari Prasad Basu mentions Bose's reprimanding his newly-found friend Sister Nivedita (1867–1911) at about this time about her casual attitude to violence. She had told Bose of her plan to join a group of *shikaris*. This prompted Bose to hold forth on the cruelties of *shikar*. Basu asks rhetorically: 'How old was this disgust with the cruelty of *shikar* in Dr Bose?'[53] Had Basu been more cynical, he could have pointed out that Bose probably delivered this reprimand in a house decorated with a large number of stuffed animals which he had himself shot.

These reasons were probably held together by a deeper motivating force. There is a clue to this in a letter to a friend written in the following year:

> I hear, from time to time, a call from the mother. I as her servant must start by collecting the dust of her feet as a benediction. You and all my friends must bless me, so that this servant can serve the mother with all his heart and soul; and his strength for work can increase.[54]

We shall return to the implications of this 'call' later. This change in the direction of Bose's research was also brought about by his deepening relationship with an Irishwoman named Margaret Noble who adopted the name Nivedita after she became a disciple of Swami Vivekananda and came to India in 1898.[55] She came as a

[51] Bose, *Abyakto*, p. 135.

[52] Ibid.

[53] Basu, *Nivedita*, vol. 1, p. 583.

[54] J. C. Bose, Letter to R. N. Tagore, 23 June; cited in Gupta, *Jagadish Chandra*, p. 35.

[55] Vivekananda was doing in religion what Bose was doing in science. Born

social worker and immediately became active in the Ramakrishna Mission established by the Swami. Bose met her soon after her arrival and it was 'friendship at first sight'. She was then thirty-one and Bose forty.

The friendship between the nun and the scientist gradually turned into a deep platonic bond. Those were difficult days for Bose. He was not yet well known enough to have his way with the authorities, and the absence of proper research facilities at the Presidency College was beginning to demoralize him. Nivedita's strong yet supportive personality and her burning faith in a science that would reflect Indian sensitivities must have given him a new faith in his work. Bose, being a Brahmo, still did not have much patience with Nivedita's—and the Ramakrishna Mission's—version of Hinduism. But this was a minor irritant in a relationship which was quickly becoming central to Bose's life. Three developments contributed to this.

First, in 1898, Nivedita introduced Bose to Mrs Ole Bull, an American friend of India. She encouraged Bose to address Mrs Bull as 'mother' and induced Mrs Bull to accept Bose as her son, as she herself had already done metaphorically by calling him her 'bairn'. One of the major pay-offs of the Bull–Bose relationship, which Nivedita nurtured till her death, was the financial help Mrs Bull gave to Bose's research.[56]

Second, Nivedita, searching for evidence of Indian greatness in Vedantic Hinduism, could not but stumble upon the distinctive Indian concept of vitalistic, organic monism, if that is the right expression, that was implicit in some of Bose's work. Almost as soon as she came to know Bose, Nivedita took over the responsibility of editing his work.[57] Bose was neither the first nor the last

Narendra Nath Dutta in 1863 in Calcutta, in social circumstances roughly similar to those in which Bose was born, Vivekananda was India's first modern and nationalist Swami and missionary of Hinduism to the West. He died in 1902 when Bose was at the pinnacle of his glory.

[56] Actually, after Mrs Bull's death, her daughter began harassing Nivedita about these 'wasteful expenditures' on an Indian scientist. The other person who went with a begging bowl to rich Indians to collect money for Bose's research was Rabindra Nath Tagore. He also suffered many humiliations and attracted the hostility of other clients competing for the favours of these patrons. Particularly unpleasant was Tagore's encounter with the court of Raja Radhakishor Devanmanikya, Bose's major Indian patron.

[57] The editing was so heavy that Nivedita could legitimately be considered a junior

person to receive such editorial help from Nivedita, but he was certainly the only scientist among Nivedita's acquaintances who had something important to say and was looking for an appropriate idiom. Nivedita's basic training was in science and the areas in which Bose had specialized were, as yet, not beyond the understanding of someone with a good general education in science. She was not, therefore, shouldering an impossible task. For instance, Bose's best-known book, *Responses in the Living and the Nonliving*, reported his own researches but owed its elegant style as well as structure to Nivedita. Till she died the collaboration survived. Strange though it may appear, the first articulation of Indianized science was in the language of a western woman.

Third, after his marriage Bose began undertaking a series of long sentimental trips to what he called the relics of the glory of ancient India. The trips, often undertaken in arduous, even dangerous conditions, began to have Nivedita as their constant feature after 1899. They often lasted months and brought Bose even closer to Nivedita.

The growing closeness to Nivedita introduced some changes in Bose's life. He gradually moved away from Abala's protective umbrella in professional matters. Though in other respects he continued to depend on her, in science his new motherly protector became Nivedita. This induced in him a slight sense of guilt, too. It was at about this time that Bose started making occasional private requests to his friends to praise Abala and her social work in public.

All this subtly changed Bose's research interests. However, there was nothing subtle about the results of the change, which were, if anything, dramatic. The turning from the study of the inorganic to that of organized life, as Bose described it, led to studies of the responses made by plants, animals and the nonliving to various types of mechanical and biochemical stimuli. It meant a sudden spurt in productivity too. In the five years 1894–9, Bose had published only four papers, whereas in the three years 1899–1902, he published nine papers and a book. It is an index of the radical nature of the change that during his third and last phase

author of some of Bose's work. According to Basu, *Nivedita*, vol. 1, p. 660, during the first ten years of her friendship with Bose, Nivedita edited about 2,500 pages and prepared about 1,000 charts and diagrams for the four books which Bose wrote during the period. This is apart from the work she put in on his large number of papers.

of work, Bose was still trying to consolidate and extend the work of his second period, rather than break new ground.

The thematic continuity in Bose's work, beginning 1899, is also reflected in the evocative titles of his papers on botanical and para-botanical researches.[58] What was a new scientific idiom, and has sometimes been explained away as only an idiom, also spilled over into his other creative work. In 1920 Bose published a collection of essays in Bengali which articulated the themes predominant in his technical papers;[59] in fact, some of the essays were actually attempts to popularize the themes.

The core of Bose's research interests was now the similarities between the living and the nonliving and a biological model that would explain physical phenomena. In 1901, for instance, he wrote:

> I have invented an instrument in which any pulsation or response created by pinching would be recorded by itself.... And just as you feel the throb of life by feeling the pulse, similarly the throb of life in the inanimate object is recorded in my instrument. I am sending to you a very astonishing record. Please observe the normal coursing of the pulse, and then how it moves under the effect of poison. The poison was applied on an inanimate object.[60]

The credo of such work was roughly this:

> There is no break in the life-processes which characterize both the animate and the inanimate world. It is difficult to draw a line between these two aspects of life. It is of course possible to delineate a number of imaginary differences, as it is possible to find out similarities in terms of certain other general criteria. The latter approach is justified by the natural tendency of science towards seeking unity in diversity.[61]

[58] A random sample: 'How the Plants React to Pain and Pleasure' (1915); 'Testing the Sensibility of Plants' (1915); 'The Unity of Life' (1927); 'Is the Plant a Sentient Being?' (1929); 'Are Plants Like Animals?' (1931); 'Injured Plants', and 'Inorganic and Organic Memory'. For a complete list see S. Bala Subramanian, 'Bibliography of Books on and by Jagadish Chandra Bose' (Calcutta: National Library, 1965), mimeographed. See also the chronology in Home, *Acharya Jagadis*, pp. 75–82.

[59] Some examples are 'Birth and Death of Plants', 'Literature in Science', 'The Mute Life', 'Injured Plants', and 'Gestures of Trees'. All the essays have been reprinted in Bose, *Abyakto*.

[60] Part of a letter probably written to Lord Rayleigh or Sir James Dewer, 3 May 1901, cited in Gupta, *Jagadish Chandra*, pp. 41–2.

[61] Bose, *Abyakto*, p. 87.

The statement underlines the paradox that was Bose. While it seems so contemporary as a statement of a personal philosophy of science, it sought to legitimize an approach which in its specifics was to prove simple-minded in the context of plant physiology within a short time.

The reaction of western scientists to such work was mixed. Some found the approach magical and mystical, and hence worthless; others found it appealing for that very reason. When Bose demonstrated, at the Royal Institute in 1901, the death agony of a poisoned tinfoil, and cured another with drugs, he merely took his biological model of physical phenomena, as both groups had expected, to its logical conclusion. After his peroration, Robert Austen, the greatest living authority of the time on metals, 'was beside himself with joy'. He reportedly said:

> I have all my life studied the properties of metals. I am happy to think that they have life.... Can you tell me whether there is future life—what will become of me after my body dies?[62]

The culture of a science may sometimes serve as a good projective medium for individual scientists and even individual societies. But the structure of scientific knowledge never has that special weakness for any specific metaphysics or for the needs of cultural nationalism. Nor, for that matter, can it be expected to cure the anxieties of scientists at the level at which Professor Austen so innocently articulated them. As it happens, more sophisticated, though duller, interpretations of some of Bose's data are today available. Even in his lifetime, despite their apparent esotericism, the salient features of Bose's discoveries were being gradually integrated within the framework of formal science.[63] Bose, however, was never open to alternative explanations of his work. As with many pioneers, he had a hard core of dogmatism hidden by an other-worldly style. He was the obstinate mother's son who never gave in and—one must understand—could not give in. Like his ideal teacher, Ishwarchandra Vidyasagar (1820–91) and his ideal of divine creativity, Shiva, Bose felt justified in

[62] Gupta, *Jagadish Chandra*, p. 44. Austen may have sensed that Bose's work was, among other things, a remedy for—or defence against—the fear of death. See section VII for an analysis of how Bose utilized the universal human anxiety about death to organize his identity.

[63] See D. M. Bose, 'Jagadish Chandra Bose', pp. v–xv, for a brief account.

combining softness with obstinacy, protectiveness with rage.[64] For he had, he was sure, found his certitudes. His impatience with criticism was a by-word among his students and friends. And during the last twenty-five years of his life, he went on redefining and sharpening his concepts without ever substantially modifying them.

It is the middle-aged scientist that many remember. Some of his students remember him as a stout, short, impeccably dressed, energetic man—intolerably narcissistic, impatient, abstruse, and authoritarian. Satyendra Nath Bose, the most gifted and most outspoken of them, remembers him as an essentially selfish, ruthless old man who exploited rather than expressed nativistic themes. The students also remember Bose as a good orator and conversationalist, in spite of his East Bengali accent. This estimate, however, Bose would not have accepted.[65] His distrust of his orality was deep. Even in his old age, he spoke as little as he could and, when he had to lecture, took days to prepare for it. So much so that many remember his public lecture-demonstrations essentially as well-rehearsed dramatic performances, in which nothing ever went wrong: 'He would even use the same set of bricks to heighten his table every time', one of his students affirms. Abala knew this. In his last years it was she who often pressed Bose into giving his customary annual lecture to the Bose Institute.

Bose was particularly aware of his own oral aggression. Once, while appearing before the Royal Services Commission, he took along with him a memo exhorting himself: 'Do not get angry.'[66] The clearest indicator of his problems with the management of oral aggression was, however, his tendency to stammer when provoked or excited. Time has wiped out most details of the trait, but it is not impossible to guess its meaning in a person such as Bose. A classic clinical text mentions the stutterer's 'hostile or sadistic tendency to destroy his opponent by means of words', so that stuttering

---

[64] Bhattacharya, 'Acharyadeb Smarane'; Gupta, *Jagadish Chandra*, p. 94; and Bose, *Abyakto*, p. 220d. Ishwar Chandra Vidyasagar (1820–91) was an eminent Sanskritist, social reformer, and teacher. His aggressiveness, directed mostly against impersonal sources of injustice and ignorance, could not really be compared with Bose's free-floating aggression. The idealization of Vidyasagar and Shiva were rather a search by Bose for a rationalizing principle that would make his anger more acceptable to him.

[65] Gupta, *Jagadish Chandra*, p. 79.

[66] Rabindra Nath Tagore seemed to agree with Bose. See his *Chithipatra*, ed. P. B. Sen (Calcutta: Viswabharati, 1957), vol. 6, p. 107.

becomes 'both the blocking of and the punishment for this tendency.' The stutterer's hesitation often expresses his desire to kill, 'because in him speech remobilizes the infantile state when words were omnipotent.' He is a warrior who is perpetually trying to control his weapon.[67] For Bamasundari's son and one with such mixed feelings about his cultural origins and status, though, such functional disorders could symptomatize other ambivalences too:

> Occasionally, the words that should or should not be uttered have, on the deepest level, the significance of introjected objects.... The stutterer may not only unconsciously attempt to kill by means of words...his symptom also expresses the tendency to kill his words, as representing introjected objects.[68]

There was also Bose's ugly manner in classroom and laboratory. At the slightest provocation he would shout at his students and colleagues 'moron' or 'I shall whip you' ('*chabke debo*') or 'you are dismissed'. Frequently he felt sorry after his outburst and apologized, but this was hardly adequate compensation for his colleagues, already sensitized by Bose's inelegant patriarchal style. Many of them left him as his sworn enemies.

More than one of his students claim that Bose was less short-tempered when he was younger. Perhaps his aggression which in his more creative days was temporarily turned inwards had turned outwards again.[69] Perhaps the restraint once exercised by his identification with his father was now less effective. Whatever the reason, his abrasive autocratic style became in his middle years one of his major identifiers. At least one informant relates this drift towards authoritarianism to Bose's increasingly greater exposure to the *guru-sishya* concepts popularized by the Ramakrishna Mission, particularly Vivekananda and Nivedita. He now cultivated and increasingly enjoyed his public image of a distant *acharya*—a religious teacher cum preceptor. Certainly this was the image Nivedita wanted him to present to the world.

The fulcrum of the Bose–Nivedita relationship also subtly shifted. For a long time, Bose had had a Brahmo's reservation about Nivedita's faith. In a letter he had once affectionately but

[67] Otto Fenichel, *The Psychoanalytic Theory of Neurosis* (Now York: Norton, 1945), pp. 312–13.

[68] Ibid., p. 314.

[69] See the comparable dynamic in Isaac Newton in Frank E. Manuel, 'Newton as Autocrat of Science', *Daedalus*, 1968, 97, pp. 287–319.

belligerently attacked her for her support to the caste system, Kali worship, etc.[70] On the other hand, Nivedita had never cared much for Bose's Brahmoism; nor did she take it very seriously. She had always hoped that some day Bose would move closer to her concept of Hinduism,[71] but apparently, both remained steadfast in their religious ideology till the end of their days. But while Nivedita stuck to her beliefs, Bose in fact changed under her influence in his middle years. According to Geddes, he became a celibate, which was a major plank as well as value of the Ramakrishna Mission group and which subtly balanced the scales between Abala and Nivedita.[72] Bose also physically moved close to the Advaita Ashram, Mayabati, in the foot-hills of the Himalaya. At Mayabati, he claimed, ideas rushed into his head whereas in Calcutta they dried up.[73] Young, self-confident Vivekananda had once perceptively commented to Nivedita:

> Yet that boy [Bose] almost worshipped me for 3 days—in a week's time he would be my man.... And those are always the people who make the fuss about worship of the personal. They don't understand themselves, and they hate in others what they know they are struggling against.[74]

The Swami may have been wrong in his prediction; he was not wrong in his analysis.

Bose's growing authoritarianism was also associated with diminishing self-esteem, a weakened capacity to handle professional criticism, and a growing inability to admit that he could be less than fully autonomous as a scientist. For instance, his publications gained immensely from Nivedita's editorial efforts: of Nivedita's few available letters to Bose, at least two refer to Bose's works as 'our books'. Yet, except in a stray letter he never expected to be published, Bose did not openly acknowledge Nivedita's contribution. Though he paid handsome tributes to his friend at various places, nowhere did he mention, not even in the preface to his books, the actual nature of Nivedita's help. On the contrary, he and some members of his family were reluctant to publish letters from Nivedita in which the westerner innocently mentioned the

[70] Basu, *Nivedita, 1*, pp. 594–5.
[71] Ibid., esp. pp. 585, 592n.
[72] Geddes, *Life and Works*, Ch. 17.
[73] Basu, *Nivedita*, vol. 1, pp. 594–5.
[74] Ibid., third photographic plate facing p. 592.

details of her scholarly assistance to the East.[75] Similarly, there was Bose's defensiveness about the criticisms of the famous botanist Sidney Vines. Vines had been his teacher at Cambridge and Bose often sent him his drafts for comments. But he would later carefully erase his pencilled comments, so that nobody knew the snide remarks which Vines, always considered a supporter of Bose, had in fact made about some of Bose's later work.

Nothing sealed Bose's scientific fate more completely than his inability to keep open his channels of communication with his colleagues. He never allowed anyone to contradict him in either academic or non-academic matters; he could not enjoy the company of young scientists or, for that matter, of anybody whom he considered lower in social status (even though he was considered 'jolly' company by his 'equals'); and he refused to see the experimental results his colleagues obtained if the results contradicted his own ideas. Increasingly he used his collaborators as laboratory assistants. They did the experiments and collected empirical data, having no say whatsoever in decisions regarding theory, methods, and interpretation of results. Every morning they had to attend an informal but strict roll call, after which they were assigned work for the day by their stern taskmaster. No wonder that when some scientists like R. Snow began to offer less vitalistic interpretations of some of Bose's own experimental results, he rejected their work out-of-hand.

Simultaneously, Bose worked hard to maintain the image of a humourless, puritanic Brahmo among his assistants. He was secretive about his reading habits, particularly his fondness for the best-selling middlebrow novels of Saratchandra Chattopadhyay and the humorous short stories of Rajsekhar Bose. He even cut

[75] The instrumental use made by Bose of Nivedita is also evidenced by her articulating some of his tall claims. For instance, she took Bose seriously when he said to her that he was the first to send wireless messages and that this was not admitted because of racism. This was certainly not true because wireless messages had been sent earlier in England. However, Bose had done excellent work in the area and had demonstrated his wireless equipment before G. M. Marconi did.

Incidentally, many of Bose's claims in the area of plant physiology were later to embarrass scientists at the Bose Institute after his death. A typical example was the evidence Bose claimed to have in support of his theory of a pulsating or valve activity in plants to account for the ascent of sap. He advanced this nervous theory against the theory of osmotic pressure. See also Bose's revealing letter (dated 7 October 1937) to Mr Herbert, who was planning to co-author a biography of Nivedita in Basu, *Nivedita*, vol. 1, plates 16 and 17 facing p. 592.

down on smoking, in keeping with the new puritanic posture, though his declining health too had something to do with this. He also now began to take care to hide his curiosity. He had always been an inquisitive, adventurous person as a child, but in middle age he came to hate anyone with similar qualities.

In all, one gets the impression of a harried man trying desperately to contain his feelings of inadequacy by affirming his power and uniqueness, and by posing as a larger-than-life figure. Those suspicious of the sudden emergence of well-organized character traits may see in these the evidences of a deeper self. In giving up the earlier egalitarian bonhomie with students and assistants, they may say, Bose was being true to a part of his self that had become important for him in the course of years. But by all accounts, the change was a painful process for his friends.

There were other indices, too, of the disintegration of his well-organized psychological defences. Bose had become by this time a compulsive penny pincher. He often personally visited the suppliers of his laboratory and wasted hours bargaining for small discounts. He refused to trust juniors or delegate authority in any financial transaction involving his research team. Bose's niggardliness, however, had its comic side. Overly fussy about money, he often shocked students by his gullibility. He mostly bargained for small cuts, more symbolic than substantive, and was pacified if he could convince himself that he had got a better deal than his juniors could have.

By this time he had redecorated his house in fully Indian style and given pride of place in his living room to a famous painting of Mother India or Bengal—like all true-blooded Bengalis, Bose did not, or could not, distinguish between the two—by Abanindranath Tagore. Patterned after the image of Durga, the traditional mother goddess of Bengal, the painting was another Brahmo's homage to the core archetype of his culture, one which had presided over Bose's work all through his life. He had also become obsessively finicky about food, apparently because of diabetes, blood pressure, and a 'nervous breakdown' in 1915.[76] But even earlier, his growing

[76] One wishes that something more was known about the breakdown; how for instance Bose's body, if it *was* a matter of the body, spoke the language of his inner struggles and what, other than the strain of an international journey, precipitated the symptoms. Only one biographical account provides enough material for us to guess that Bose was in an acute anxiety state and that it was at least partly a result of constantly preparing and delivering public lectures on his work during the trip. See S. C. Ganguli, *Acharya Jagadis Chandra* (Calcutta: Shri Bhoomi, pp. 141–3.) Apparently, in sickness as in health, Bose's anxieties centring around orality remained the central dynamic of his personality.

rejection of food was obvious to many (part of Abala's heroic effort towards an integrated Bose was that after her marriage she learned to cook very well).

For a short time, as a young man, Bose had a large circle of friends but, as the years went by, the radius of his interpersonal world shrank. As he grew less friendly towards his students, he distanced himself from his close friends, too. Nowhere was Bose's insecure self-image and almost paranoid fear of hostility and rejection by others as obvious as in the case of Rabindranath Tagore. The fate of this relationship deserves, therefore, a brief digression.

I have already mentioned Tagore's early support to Bose, both psychological and financial (through the patronage of the Maharaja of Tripura, Radhakishor Dev Manikya). For this, the poet even had to face the jealousy and hostility of some other protégés of the Maharaja. In other ways too, Tagore, as the acknowledged leader of Indian intellectuals, had been warm and protective towards the scientist. But the middle-aged Bose began to take Tagore's help for granted and bear him a grudge for not doing enough. He avenged this 'neglect' by refusing to help the poet in any way when the latter was trying to establish a university at Shantiniketan. And when Tagore won the Nobel Prize for literature in 1913, his friend was less than enthusiastic. Though Tagore publicly pleaded for greater recognition of Bose's work in the West and proposed him for a Nobel Prize, Bose felt unjustly neglected. His achievement concerns were always tainted by jealousy and he could never fully master his basic feelings of rejection. Moreover, he may have perceived—with the shrewdness which only a man with strong paranoid feelings is capable of—that Tagore now cared less about his science and, despite being a littérateur, that he had sensed that the future did not belong to Bose's science.

Bose now seemed to be looking for any excuse to break away from a man he had once declared was one of his best friends. The break came when Bose's nephew Aurobindo, whom Bose had himself sent to Shantiniketan, became greatly attached to Tagore and his institution. Bose was particularly fond of this nephew and possessive about him in a way only a childless man can be about a boy who would never be entirely his own.[77] He now began to fear

[77] D. M. Bose, his other nephew, mentions that in his childhood he was afraid of the enveloping, and slightly suffocating love of his uncle. Sudeb Rai Chaudhuri, '*Byakti o Byakitva*', *Desh*, 1973, *40*, pp. 751–5.

that he would lose Aurobindo to Tagore. He did his best to wean the nephew away from Shantiniketan and began a sustained but low-key private campaign against the poet among his acquaintances. Though ostensibly the poet and scientist maintained their friendship, by now both were wary of each other. It must have been singularly hard on Tagore who, out of a sense of patriotic duty, had virtually gate-crashed into Bose's life, to help him with money, encouragement, and 'propaganda'. The bitter loneliness in which he faced Bose's rejection, at a time when such personal differences could not be aired in public, can be gauged from the Bengali obituary of Bose he wrote in 1937.[78]

Gradually, Bose cooled towards Prafulla Chandra Roy (1861–1944), the illustrious professor of chemistry, and quarrelled with Devmanikya, his erstwhile patron. He also began to avoid participating in movements, large organizations, and public functions.[79] Both Mrs Bull and Nivedita were dead and recognition from the West, never complete, now became more uncertain. He had committed enemies there and his weakening grasp on his discipline made their task easier. After his mental illness, even his physical links with the West and westerners became weak.

These centripetal emotional forces may have been by-products of Bose's greater involvement in research and his attempts to find steady financial support for his work from the government, but they could also be the reflections of something deeper. He had made new and more reliable friends by then:

> In time, plants came to be his best friends. He loved them, reared them and treated them with tender care. He followed their life history and perhaps they also spoke to him. Pain and relief from pain in the plant became clear to him.[80]

No account of Bose's life can be complete without a word on the advanced research centre he established in Calcutta in 1917. The Bose Institute was an ornate, temple-like structure meant 'to search for the ultimate unity which permeates the universal order and cuts across the animal, plant and inanimate lives.'[81] Bose played a large part in the design of the building, being greatly influenced by

[78] Tagore, *Chithipatra*, pp. 6, 124–8, esp. p. 128.

[79] Roy and Bhattacharya, *Acharya*, pp. 231–2.

[80] Gupta, *Jagadish Chandra*, p. 92. Roy and Bhattacharya (*Acharya*, p. 216) also report that Bose's neighbours believed that they saw the scientist conversing with plants at night.

[81] From the inaugural speech reproduced in Bose, *Abyakto*, pp. 142–58.

Ajanta and Ellora. It is said that Nivedita before her death had discussed some of these architectural possibilities with Bose. Called Basu Vigyan Mandir (the Bose Temple of Science), the Institute had a special platform or *vedi* for its founder to sit and meditate on. In its garden were deer, peacocks, and 'talking birds'. Near the entrance to the Institute was a sculpted relief of Nivedita who had hoped and planned with Bose for such an institute and whose unconditional adoration had, to a great extent, made Bose what he was. Under the relief, Bose secretly buried a small box containing Nivedita's ashes. As emblems of the temple, Bose selected sculptured representations of the Sun-God (a symbolic pointer to the identification that had been his first bridge between *bhakti* or devotion, and knowledge), the *vajra* or thunderbolt (the weapon with which Indra fought evil in the form of demons, and a traditional symbol of legitimate fury), and *ardha amlaki*, the Buddhist symbol of total renunciation. Some of these symbols were said to be Nivedita's suggestions.

Bose's attitude to the Institute and to science as a vocation is revealed in the life-long commitment he expected from those working there. Many of his colleagues promised—or rather, if one takes into account the psychological pressures in the institution, had to promise—in writing that they would never leave the Institute.[82] Bose himself admired austerity, but never cared to live by it,[83] and yet expected total austerity in his subordinates. He also came to believe that only East Bengalis could give him the allegiance he wanted. He gave preference to them in recruitment and trusted only them in the affairs of the Institute. But his peculiar mix of self-hatred and violence affected even his treatment of them. Though he himself had a pronounced East Bengali accent, he was bitterly sarcastic towards any one in the Institute whose accent evinced an East Bengali origin.

However, Bose had the authoritarian man's special capacity to elicit and hold the loyalty of some of his subordinates. So when he began to stretch his experimental results and force his associates to do the same, presumably in the interest of the higher science he was striving towards, nobody publicly protested. And the private criticism of these associates survived only as memories of the deviation from principles and opportunism which a revolutionary

[82] The promise at first had to be given on a piece of paper. Later on, it used to be engraved on a copper plate.

[83] Geddes, *Life and Works*, p. 220.

movement cannot always avoid. For many those were the most glorious moments of their lives, and it seemed better to participate and occasionally fail than to be a mere onlooker.

The major grants towards the Institute's buildings and its research programmes came from the British Indian government. Nivedita had died in 1911 and the conduits for financing research that she had built with such care were now dry. And Bose had by then quarrelled with his Indian patrons. Fortunately, he had maintained friendly relations with the political authorities for many years. The imperial system then was run by self-confident administrators, well versed in the management of colonial societies and not prone to overreact to Bose's brand of nationalist science. Also, it had not gone unnoticed that Bose had always been in his personal life fairly attentive towards westerners and not very accessible to Indians. (Bose's nationalism was not merely tinged with a sense of inferiority, but also with a sharp awareness of where the metropolitan centres of science were located. One of his assistants mentions that he expected Abala to relate to his Indian visitors, the western scientists' eulogies of his work. If Abala, a more self-confident and autonomous person, forgot, Bose would take care to remind her of her duty.)

Some of his contemporaries found such links unacceptable. Prafulla Chandra Roy, an ardent nationalist, privately called Bose a *dhamadhara* or a sycophant.[84] Satyendra Nath Bose, the physicist, was harsher: Jagadis Chandra, he said, was an 'utterly selfish' man who was not merely careful to be on the right side of the British government, but also had enough political sense to build the right type of rapport with the highest authorities without hobnobbing with junior officials. Both critics failed to understand that Bose's nationalism was as real as his emotional investment in a West which was close to him by virtue of being his counterplayer; psychologically, he could not do without the West. He had after all been anti-imperialist when it was troublesome to be one—during the 1905 movement against the partition of Bengal. If he was ambivalent towards the West, it was the ambivalence of *babu* nationalism itself, a nationalism arising in response to western intrusion and binding Indians to their counterculture in a love–hate relationship. In the case of functioning scientists, the ambivalence

[84] Ray himself, however, had gone bankrupt as a researcher in middle age, trying to be a 'nationalist' chemist and becoming, in the process, mainly a patron of young scientists.

had to be deeper. The absence of knowledgeable consumers of modern scientific research in India forced them to look westwards to find such consumers and protect their self-image as scientists. Their country respected them mainly as grand successes without appreciating the content of their work. Even this success was defined in terms of bureaucratic status, degrees, wealth, and the opinion of established scientists outside India. It is to Bose's credit that despite these circumstances, he was one of the first to suggest a definition of Indianness in science and to try to integrate in the definition the role of the modern scientist. It is less important that he could not fully consolidate these gains.

But he tried all the same. Dedicating the Bose Institute 'to the feet of God for bringing honour to India and happiness to the world', the aging patriarch said:

> The excessive specialization in modern science has led to the danger of losing sight of the fundamental fact that there can be but one truth.... India through her habit of mind is peculiarly fitted to realize the idea of unity, and to see in the phenomenal world an orderly universe. It was this trend of thought that led me unconsciously to the dividing frontiers of the different sciences and shaped the course of my work in its constant alternations between the theoretical and the practical, from the investigation of the organic world to that of organized life and its multifarious activities of growth, of movement, and even of sensation.
>
> What I established today is a temple, not merely a laboratory. Truths which can be sensed are determined by experiments; but there are some great truths which can be reached only through faith.[85]

Bose died in 1937, a renowned scientist and a venerated academic, surrounded by loyal disciples. But neither the fame nor the loyalty could mask his alienation from some of the newer currents of scientific thought. Even before his death, his professional defeat was apparent, if not to him, at least to his students and co-workers, monitoring and occasionally fudging data for him. Some of these colleagues had also become secretly sympathetic to the increasingly convincing technical arguments and experimental data many non-vitalistic plant physiologists were marshalling against Bose.

This professional isolation was not merely due to Bose's

[85] For the full speech, see 'Nibedan' in Bose, *Abyakto*, pp. 142–58. The first paragraph is an English translation taken from Gupta, *Jagadish Chandra*, p. 134.

advancing age, a handicap which can be fatal in any fast-growing discipline. It was also due to three changes in the environment of science. First, as the memories of World War I faded, the first shock of seeing the other face of modern science subsided. This reduced the appeal of Bose's humane, pacifist, eco-sensitive science for other scientists. Secondly, Bose's monistic philosophy of science, which anticipated many of the concerns of the contemporary philosophy of science, was for his own time a premature development within the dominant culture of science. The manipulative, power-seeking, narrowly empirical and operationalist aspects of the dominant culture were yet to be threatened by the growing self-confidence of the critical social theories of science. Thirdly, there was a change in the state of the disciplines within which Bose worked. With the consolidation of new paradigms, these disciplines reverted to their unheroic normal activities.[86] And though they did not bypass Bose, they began to 'undramatize' his discoveries in a way which he, had he cared to know, would have found heartless in the typical western way.

VI

When science is universal, can there be in the world of science a place which will remain vacant without an Indian devotee? There certainly is.[87]

To understand the identity Bose evolved as a creative Indian scientist, one must take into account three recurrent themes in his life and work. The most salient and self-conscious of these themes was pithily summarized by Bose himself as a simple principle:

> The consciousness of the scientist and the poet both go out in the search of the inexpressible one. The difference lies in that the poet ignores the means, the scientist does not.[88]

If this was an off-the-cuff reflection on his role as a scientist, the concluding remarks of the lecture-demonstration he once gave at the Royal Institute were a clearer proclamation of his professional identity as it mediated his self-concept and his concept of true Indianness:

[86] The concept of normality is of course borrowed from T. S. Kuhn, *The Structure of Scientific Revolution* (Chicago: University of Chicago, 1962).

[87] Bose, *Abyakto*, p. 150.

[88] Ibid., p. 87.

> I have shown you this evening the autographic records of the stress and strain in both the living and the non-living. How similar are the two sets of writing, so similar indeed that you cannot tell them one from the other? They show you the waxing and waning pulsations of life—and climax due to stimulants, the gradual decline in fatigue, the rapid setting in of death rigour from the toxic effect of poison. It was when I came on this mute witness of life and saw an all-pervading unity that binds together all things—it was then that for the first time I understood the message proclaimed on the banks of the Ganges thirty centuries ago—'they who behold the One, in all the changing manifoldness of the universe, unto them belongs eternal truth, unto none else, unto none else.'[89]

This must have been hard to swallow for some of his western listeners, steeped in the positivist culture of science. But it electrified the Indian élites. In this resolute attempt to obliterate the differences between the world and the self, and the living and the dead, they saw a convincing synthesis of the foreign and the indigenous. And in this synthesis, the indigenous ideals of knowledge and inquiry predominated. It not only helped them to accept their Indianness which colonialism had made a controversial inheritance, but also helped resolve their ambivalence towards western science.

For centuries the Indian imagination has used nondualist thought to impose order on diversities, contradictions, and oppositions, and a unified worldview on a fragmented society. The institutional emphasis in India on social hierarchy, plurality of norms, and segmentation of interpersonal relations has paradoxically underlined the ideological stress on the oneness of existence and the singleness of experience. The parallel attempt to tame individualism, by instilling the awareness of the unity between *brahman* (essence) and *atman* (essential reality of self) has often been a source of creativity in the Indian theory of knowledge and a rationalization of dissent.[90] The nondualist concept of an impersonal, timeless absolute gives a special meaning to the concept of scholarship. The scholar is expected to extend the perimeters of empirical knowledge while being open to its transcendental

[89] R. N. Tagore, 'Acharya Jagadisher Jaivarta', *Vasudhara*, 1957, vol. 2, pp. 107–9.

[90] At least one psychologist has felt this to be associated with easier acceptance of the new, the strange, and the different. G. Murphy, *In the Minds of Men* (New York: Basic Books, 1953), pp. 44, 268.

meaning. The idealist strands of Indian scholarship derive their primacy partly from this particular construction of the links between knowledge and reality.

Bose formulated this philosophical position into a scientific idiom and a research ideology. Once he did this, a number of compatible concepts of true knowledge, the unity of science, inter-disciplinary method, and the social obligations of a scientist became available to him. Through this culturally tinged set of concepts he constructed a positive self-image in the fluid scientific culture, generated culturally valid and personally meaningful symbols, and ensured his survival in the often-hostile Cartesian scientific culture of the modern West.

The ideology helped Bose cope with deeper diversities, too. The 'multiform unity in a single ocean of being' and 'the great pulse that beats through the universe' came to represent a personality integration he had constantly struggled towards.[91] Nothing tells the story better than a recurring motif of his life: a personal myth which became, in two senses, a concrete design. The beginning of this motif lay in Bose's early fascination with a rivulet that, springing from a hillock, flowed out from under a culvert near his childhood home and eventually merged with the Padma, the majestic easternmost branch of the Ganges, flowing towards the Bay of Bengal. This fascination with the theme of originating at an elevation, a temperamental movement or journey and final merger with a calmer but grander oneness persisted through his life. When at St Xavier's School, he made a small garden in his hostel within which was an artificial brook spanned by a bridge. Again, in two of his houses and in the Bose Institute, Bose designed artificial streams originating at a height (in a fountain, for instance) and flowing under bridges into the placidity of larger rivers. In his Darjeeling house—he called it *Mayapuri* or enchanted place, and it was reportedly the only place where he was not racked by insomnia—he did not create the motif. But he saw to it that the house was situated between two waterfalls from which two rivulets flowed noisily down to their own quieter selves a few hundred feet below.

The most famous articulation of the motif, however, was in a brief travelogue, published as an essay in 1894, the year he took the vow of fidelity to research.[92] The essay, highly personal and

[91] J. C. Bose in Ramananda Chatterji (ed.), *The Golden Book of Tagore* (Calcutta: Golden Book Committee, 1931), p. 16.

[92] Bose, *Abyakto*, pp. 73–81. For the Bose Institute he also commissioned a mural

elegantly poetic, describes his search for the origin of Bhagirathi or Ganges in a sacred fountain in the Himalaya, the shaky rope bridge over it near its source, its turbulent journey towards the plains, and its final transformation into the peaceful, creative, divine motherliness of the Ganges of the plains. He rediscovers the critical origin in the snaky, matted plaits (*jata*) of Mahadev or Shiva—the god of destruction, the mythic consort of the Mother of the Universe, and the traditional personification of phallic creativity. In the hair of the world-renouncing, austere god, Bose found the source of all life, all movement, all progress.

This fantasy of birth, beginning, and creativity, sometimes invented and sometimes discovered, had two axes: a male principle or its phallic representation as the ultimate passive, steady source of creation, creative by virtue of being able to contain its destructiveness through austerities and an active, temperamental, everchanging, feminine principle (*prakriti*) split into two.[93] The split was between a tempestuous, abrasively mobile, motive force—framed in this instance by identification with what one can only describe as a repudiated mother—and, beyond it, a secure, certain, oceanic peace, representing an ideal self and a symbiosis with the other mother within.

There was also the symbol of intervention through science in the form of the omnipresent bridge, rising above and binding 'nature', to negate or control her impulsiveness and unpredictable natural abandon. It was as if Bose had to choose his self-definition and intervene in the world while standing between 'the sly, cruel, crushing ruthlessness and the meditative affection of mother nature', two images that he knew to be 'merely the projections of one's own mind'.[94]

The fantasy of a grand fusion cannot but conceal beneath it some doubts about unity and some fears of fragmentation. Here was a person with a record of early diffusion of identifications which in youth became a protracted crisis of identity; and here was a fantasy that communicated transcendence, unity, and fusion,

---

which bore some relationship to the basic motif. It depicted, at his suggesion, the 'idol of knowledge' floating down the Ganges, with the eternal woman beside him—a representation, as a contemporary journal pointed out, of '*shakti* inspiring *purusha*'.

[93] Obviously this was compatible with the traditional Indian concept of *purusha*, the ultimate male, as a passive creator of the universe, and *prakriti* as the active participant in the process of creation.

[94] Bose, *Abyakto*, pp. 3–4.

triumphing over wild abandon and division. One suspects that the fantasy did tie together a conflicting set of identifications to tell the story of Bose's life as Bose would have liked to tell it.

Brahmoism endorsed the pattern. In it there was a built-in sanction for using *advaita* as a means of legitimizing the novel and the innovative. Rammohun Roy (1772–1833), the first Brahmo, had already shown modern India how to integrate exogenous cultural elements into Indian society with the help of Upanishadic nondualism. The technology could have been only further endorsed in a Brahmo family headed by a devout Vedantist who perhaps had sought in his new faith an antidote for his fragmented selves. The wider subcultural strain became an immediate, intimate actuality for Bose in his father. Bhagwanchandra must have been the first to combine, for himself and his son, the theory of nature with the theory of living.

However, the Brahmo concern with 'cosmic unity' was not exactly the pure monism of Vedanta. It included a conspicuous monotheistic element that showed the influence of Christianity and Islam.[95] Such a philosophy of life, by itself, could not have underwritten the 'implicit pantheism' of Bose. That pantheism had to come from the projective system of the little cultures of India—from the rituals, myths, magic, folklore, sayings, theories of nature, interpretations of ill-health, demonology, and so on. Here the abstract, universalist nondualism of the greater Sanskritic culture became a pan-psychic tendency to see the world as a living organism where the natural entities were not only endowed with life, but with the ability to manipulate human behaviour and fate. Bose's friends Vivekananda and Nivedita understood this much better than did the Brahmo leadership. Ideologically the former were more passionately committed to the philosophy of Vedanta, but in practice they constantly invoked the little culture's more dualist anthropomorphism. Understandably, their idiom reached beyond the perimeters of urban, westernized, upper-caste Bengal whereas Brahmoism slowly drowned in the blue blood of its followers. To the urban—and urbane—Brahmos, Vedanta always remained a distant Brahminical standard, legitimizing the new and the strange. The idea that it simultaneously existed as an immediate folksy theme in everyday life was too remote from the Brahmo consciousness which was trying desperately to be the vanguard in

[95] Ibid.

the running battle against Hindu orthodoxy. Bose's science cut across cultural hierarchies to draw upon both traditions of world construction.[96]

Bose's nondualism linked his creative imagination to his cultural milieu at yet another plane. It provided a frame of reference within which both Indian and western intellectuals could justify or at least explain to themselves their acceptance of Bose's science. Contrary to popular belief, western contemporaries of Bose, had they offered his type of theory or used his language, would have been totally isolated professionally. But Bose thrived, and was allowed to do so, in spite of his apparent esoterism and his highly personalized idiom. In fact, to many westerners Bose's idiom justified his science. Though he managed to collect some dedicated enemies in the West, such were the latent needs of some sections of the West at the time that he became a mouthpiece of western intellectual dissent too. Let one westerner who opted for the Indian way of life, Nivedita, speak first:

> The book on responses in living and nonliving is now triumphant. I want a far greater work, such as only this Indian man of science is capable of writing on Molecular Physics, a book in which that same great Indian mind that surveyed all human knowledge in the era of the *Upanishads* and pronounced it one, shall again survey the vast accumulations of physical phenomena which the 19th century has observed and collected, and demonstrate to the empirical, machine-worshipping, gold-seeking mind of the West that these also are one—appearing as many.[97]

Romain Rolland, himself trying to link up European intellectual dissent with Indian traditions, was equally enthusiastic.

> I hail the seer; he who by the light of his poetic and spiritual insight has penetrated into the very heart of Nature.... You have wrested from plants and stones the key of their enigma; you made us hear their incessant monologue, that perpetual stream of soul, which

[96] It has been suggested that the ability to empathize through the animation of inanimates or anthropomorphization is a basic characteristic of creative individuals. It allows them to seek and establish new relationships and a wider unity among diverse objects and is compatible with a stronger sense of gestalt (Greenacre, 'The Childhood of the Artist: Libidinal Phase Development and Giftedness', pp. 47–52). What is of interest here is the level or area of environment from which the sanction for the 'sense of actual or potential organization' is sought, and how as a result the dynamic of the latter is vitally affected.

[97] Letter to R. N. Tagore, 8 April 1903, quoted in Basu, *Nivedita*, vol. 1, pp. 627–31.

> flows through all beings from the humblest to the highest.... You have boldly added to the vast domain of Indian thought a hemisphere of the Being, which the ancient intuition of your ancient sages had already recognized—these innumerable people of the vegetable and the mineral world who surround humanity.... In the course of this century India without sacrificing anything of the richness of her profound soul, of that inner world which was bequeathed to her by centuries of thought, will add thereto the intellectual weapons of Europe....[98]

Even more interesting was the comment of *The Spectator*, a publication not then known for its pro-Indian sentiment:

> The culture of thirty centuries has blossomed into a scientific brain of an order which we cannot quite duplicate in the West. He (Bose) is a prince among physiological research workers and a prophet of his age, which has brought so many new powers to life.[99]

Others backed this by a cultural theory:

> The people of the East have just the burning imagination which could extort a truth out of a mass of apparently disconnected facts; a habit of meditation without allowing the mind to dissipate itself, such as has belonged to the greater mathematicians and engineers; and a power of persistence—it is something a little different from patience—such as hardly belongs to an European.[100]

Similar panegyrics were delivered by Einstein, Shaw, Huxley, and Kropotkin. Shaw was visibly shaken when Bose demonstrated the death paroxysms of a cabbage being roasted, even as his lesser-known contemporary Massingham of *The Nation* was touched by 'the unfortunate carrot strapped to the table of an unlicensed vivisector' and by the 'feelings of so stolid a vegetable'.[101] But most revealing was the response of an anonymous westerner:

> Centuries of men may point to Bose as a conveniently identifiable point from which to date the dawn of the new thought, just as today we put our finger on Socrates when we wish to focus our

[98] Letter to J. C. Bose, September 1927, quoted in Home, *Acharya Jagadis*, pp. 71–2.
[99] Cited in Ibid., p. 13.
[100] Cited in Basu, *Nivedita*, vol. 1, pp. 573–4.
[101] Home, *Acharya Jagadis*, p. 25.

view of that new thought which inspired the West for centuries....'[102]

The acute problem of ethics, and of scientific ethics, in the inter-war years had apparently induced the West to look eastwards for a plausible alternative lifestyle—harmonious, placid, secure, and holistic. In 1919, even *The Times* of London joined the chorus:

> Whilst we in Europe were still steeped in the rude empiricism of barbaric life, the subtle Eastern had swept the whole universe into a synthesis and had seen the *one* in all its manifestations.... He is pursuing science not only for itself but for its application to the benefit of mankind.[103]

The scientists did not like to be left behind. Gottlieb Haberlandt, for instance, was, if anything, even more direct than the popular writers and the newspapers:

> We saw that there is a sleep of plants in the true sense of the term; and finally realized that a man of genius can not only *hear* the corn grow, but also see it.... In Professor Bose there lives and moves that ancient Indian spirit, which sees in every living organism a perceptive being endowed with sensitiveness.[104]

Henri Bergson also felt that while 'in Darwin's theory of natural selection, conflict is the main theme; Jagadishchandra's research has thrown light on the continuities and on the beauties of consistency in nature and life.'[105]

Evidently, some sections of the West, terrorized by the death and devastation brought about by a science gone rabid, were nostalgically looking back—through the eyes of the missionary-scientist from India—to the period in their own history when an apparently less contaminated, more humane, contemplative scientific tradition had aroused less conflict in values. Thus, like Tagore, Vivekananda, and Aurobindo, Bose too was to have his moment of glory in the West, only to be rejected after a while as counterfeit. In retrospect, the rejection was inevitable. It is mainly as a prophet that Bose was first deified and later forgotten. It was too early for anyone to realize that in a crude fashion Bose was anticipating and

[102] Cited in Home, *Acharya Jagadis*, p. 67.
[103] Ibid., p. 67.
[104] Ibid., p. 32.
[105] Quoted in Roy and Bhattacharya, *Acharya*, p. 80.

tackling problems which within two decades would begin to erode the legitimacy of modern science itself. If Bose's philosophy of science seems less dated today than the scientism of the next generation of western intellectuals, it is mainly because the status of modern science itself has changed dramatically during the past few decades.

The West needed Bose's science on another, more mundane plane. Interdisciplinary research had not yet become common in the natural and biological sciences. Nonetheless, the growth of these sciences had brought about a recognition that such fence-breaking could lead to new theoretical breakthroughs.[106] This was the awareness that prompted J. A. Thompson to say:

> It is in accordance with the genius of India that the investigator should press further towards unity than we have yet hinted at, should seek to correlate responses and memory impressions in the living with their analogues in organic matter, and should see in anticipation the lines of physics, of physiology and of psychology converging.[107]

Bose's rejection of the conventional boundaries of disciplines, therefore, reflected a wider recognition in the sciences of some of their own methodological imperatives. It is only natural that this recognition should have found its most articulate spokesman in a scientist to whom the formalized divisions of scientific disciplines were merely so many indicators of academic factionalism.[108] Thus, Bose's philosophy of science bridged the past of a society trying to institutionalize new systems of knowledge and a science trying to integrate an emerging methodology. The philosophy became the means through which a particular Indian scientist, while trying to make sense of the modern scientific tradition, could also make sense to it.

The support which Bose received from the Indian élite confirmed these aspects of his professional self-concept. This support came in various forms. The Calcutta élite, with some of whom he

[106] Perhaps one should offset against the explosive growth of science in general the stagnation in the fields of plant physiology and biophysics then. These differential growth rates might have sharpened the sense of paradigmatic crises in these disciplines and allowed Bose to emerge as a major scientific figure of the time. See also D. M. Bose, 'Jagadis Chandra Bose'.

[107] Home, *Acharya Jagadis*, p. 30.

[108] Bose, *Abyakto*, pp. 84–5. See also Bose's letter of 5 October 1900 to R. N. Tagore on this subject, quoted in Gupta, *Jagadish Chandra*, pp. 39–40.

had abiding social ties and who constituted his immediate environment, reinforced his beliefs in public and in private. The vocal among them found in him a vindication of the Indian theory of life. They may not all have been scientists, but they spoke with the confidence of persons who knew Bose could not be wrong. For example, the Goethe-like Tagore, convinced that Bose's orderly universe justified his own universalism, affirmed:

> European science is following the way of our philosophy. This is the way of unity. One of the major obstacles which science has faced in forging this unity of experience is the differences between the living and the nonliving. Even after detailed research and experimentation, scholars like Huxley could not transcend this barrier. Venturing this excuse biology has been maintaining a wide distance from physics. Acharya Jagadis has discovered the unifying bridge between the living and the nonliving with the help of electrical waves.[109]

Subhas Chandra Bose, increasingly a symbol of aggressive nationalism, said:

> Discovery of life in inorganic matter points at the later trends in your research. Your research has provided direct empirical proof of the unity which the ancient sages of India had found in the varieties of life.... The magic touch of your genius has given life to that which seemed inert and insensate, it has generated a passion for a new awakening in the history of this country.[110]

Bose's students were no less enthusiastic:

> That inanimates have life, that plants have life, that both can be tired or excited...we find mentioned in many of the ancient texts of this country.... What we merely heard, the *acharya* has shown us.[111]

Thus, Vivekananda's criticism of the worth of an English botanist who had merely studied the life-cycle of a plant while the Bengali was making 'the very flowerpots in which the plant grows respond to impulses', reveals something more than an attempt to

[109] Tagore, 'Acharya Jagadisher Jaivarta', p. 107.

[110] J. C. Bose, quoted in Roy and Bhattacharya, *Acharya*, p. II.

[111] Hiren Dutta, 1915; reported in editorial notes, Bose, *Abyakto*, p. 234. Bose was, of course, called *acharya*. The other form of address used by his disciples was *acharyadeb*, teacher-god. Bose sometimes lived up to this image the hard way. While inaugurating the Bose Institute, he wore the saffron cloak of a religious preceptor, complete with the ritual sandalwood marks and *tilak*.

jump on to Bose's bandwagon.[112] It was an admission that Bose had done for Indian nationalism what no one expected a scientist to do. He had joined the modernizing Indian élite's desperate search for a form of self-esteem that would not be devoid of cultural roots.

VII

Bose's work was also shaped by his lifelong struggle with his own aggressiveness and the anxieties they aroused. In fact, his commitment to the Upanishadic worldview may have been partly due to the connotations it carried of impulse-control, mediation and resolution of contradictions. As he tried to reconcile through science the extremes of the living and the nonliving, the alive and the dead, aggressive active life and peaceful quiet death, and human violence and its 'natural' victims, he not only made his science a reparative or restitutive affair, but articulated one of the deepest cultural concerns of his people.[113]

Within the greater Sanskritic culture, the various taboos on expressions of instinctual and, particularly, aggressive impulses coextend with a theory of creativity which conceptualizes creation as the control of his destructive self by the creator, and ideal knowledge as the cognition of absolute reality acquired through suppression of all desires. Towards the end of Bose's life and under Gandhi's influence,

[112] Gupta, *Jagadish Chandra*, p. 131. Such examples can be multiplied. The chorus included P. C. Roy, the foremost Indian chemist of the time; M. K. Gandhi; G. K. Gokhale, then the grand old man of Indian politics; Subhas Chandra Bose, then a stormy young politician; N. R. Sarkar, the country's best-known physician; R. C. Dutta, the eminent historian; Brajendra Nath Seal, the philosopher and educationist who saw in Bose the 'culmination of the unifying principles of traditional Indian thought'; Satyen Dutta, a well-known poet who versified his admiration for Bose's natural philosophy; Ramananda Chatterji, the doyen of Indian journalists; and Radhakishor Devmanikya, Bose's chief patron.

[113] Some of the early psychoanalytic writers thought of creativity as a form of restitution, which coped with guilt resulting from unconscious destructive fantasies. See Melanie Klein, 'Infantile Anxiety Situations Reflected in a Work of Art', *International Journal of Psychoanalysis*, 1929, *10*, pp. 436–44; Ella F. Sharpe, 'Certain Aspects of Sublimation and Delusion', *International Journal of Psychoanalysis*, 1930, *11*, pp. 22–3, and 'Similar and Divergent Unconscious Determinants underlying the Sublimations of Pure Arts and Pure Science', *Collected Papers in Psychoanalysis* (London: Hogarth Press, 1950), pp. 137–54; W. R. D. Fairbairn, 'Prolegomena to a Psychology of Art', *British Journal of Psychology*, 1938, *28*, pp. 228–303; Harry B. Lee, 'Projective Features of Contemplative Artistic Experience', *The American Journal of Orthopsychiatry*, 1949, *19*, pp. 101–11; and 'The Values of Order and Vitality in Art', in G. Roheim (ed.), *Psychoanalysis and the Social Sciences* (New York: International Universities, 1950), *2*, pp. 231–74.

even Indian anti-imperialism began translating these emphases on non-aggression into a new political idiom which frequently equated self-discipline and non-violence on the one hand, and pursuit of power, self-government, and creative social reform on the other.[114]

Research on socialization in India, mostly based on data from the more traditional sections of the society, has something to say on the subject. The data suggest that Indian non-violence is over-determined; it represents an attempt to cope with aggressive impulses that are on the whole untamed and 'do not have a chance to be patterned or shaped.'[115] The absence of social vents through which aggressive needs could be expressed in an anxiety-free manner, and the absence of slow training in childhood which could encourage containment of aggression rather than its total denial, seems to lead to a constant fear that ego controls would fail and that violence would break out in an unpredictable, chaotic, and irrational manner, either outwards or inwards.[116]

Extrapolated into Bose's life, these insights do explain something of his scientific idiom: the emphasis on 'poisoning', 'wounding', and—to give less-than-serious examples—'pinching' and 'tickling' of plants and metal foils. Through this idiom, the ancient 'pantheistic' theme of protecting plants and the rest of apparently inanimate nature from the cruelty of humans prone to uncontrolled aggressiveness, became another link between Bose and his society. Thus, the consumers of his research became his accomplices. They helped him to contain his moral anxiety centring around aggression by seemingly sharing it.[117] The pain of a

[114] Some of these issues have been discussed in Ashis Nandy and Sudhir Kakar, 'Culture and Personality in India' in Udai Parekh (ed.), *Research in Psychology in India* (New Delhi: Indian Council of Social Science Research, 1980), Ch. 2. On the Gandhian equation between aggression control and political self-determination, see Lloyd Rudolph and Susanne Rudolph, *The Modernity of Tradition* (Chicago: University of Chicago, 1967), part 2; and Erik H. Erikson, *Gandhi's Truth* (New York: Norton, 1969).

[115] Murphy, *In the Minds of Men*, p. 51.

[116] For instance, J. T. Hitchcock and Leigh Mintern, 'The Rajputs of Khalapur', in Beatrice Whiting (ed.), *Six Cultures* (Cambridge, Mass: Harvard University Press, 1963), pp. 206–361; Rudolph and Rudolph, *The Modernity of Tradition*, Part 2.

[117] The mechanism of enlisting the consumers of creative products as accomplices by inducing them to participate in the underlying fantasy has been discussed by Fenichel, *The Psychoanalytic Theory*, p. 703; and E. Bergler, 'Psychoanalysis of Writers and of Literary Productivity', in G. Roheim (ed.), *Psychoanalysis and the Social Sciences* (New York: International Universities, 1947), *1*, pp. 247–96.

poisoned leaf, the suffering of a metal strip injured manually and cured medically—these were the data Bose worked with and the tell-tale descriptions he employed.[118] He not only personified his subjects by equipping them with animate sensitivity, he hurt them and also wanted to protect them from being hurt.

> What is the pull that removes every difference, brings the proximate nearer, makes us forget who is our own and who is not? Compassion is the pull; only the power to sympathize can reveal the real truths in our life. The ever-tolerant plant kingdom stands immovable in front of us.... They are being hurt by various powers, but no whimper rises from the wounded. I shall describe the heart-rending history of this extremely self-controlled, silent, tearless life.[119]

At another place Bose says,

> From a man's handwriting one can guess his weakness and his fatigue; I found the same signs in the responses of a machine. It was a matter of surprise that, after rest, the machine could recoup and respond again. When a stimulating drug was administered, his power to respond increased and the administration of poison made all his responses vanish.[120]

How far was this a strategy to reach the lay public, over the heads of his professional colleagues, to mobilize popular support and to secure sanction? How far was this the corollary of a deeper faith? We shall never know for sure. Some of those who knew Bose are certain that these phrases were something more than the fragments of a conscious strategy. They feel that underlying the idiom was a deep concern with anger and cruelty, survival and death. They imply that the concern had already attained a structured form when, for example, Bose showed in childhood his deep fascination with his dacoit-servant's tales of exploitation, cruelty, suffering, and death. Even at that time Bose's sympathies lay not so much with the victims as with his friend.[121] This admiration for a man who had saved his life—from death which seemed to Bose to be ever-present throughout his life—was, one suspects, something more than a transient identification. It expressed a concern that

[118] For examples see *Abyakto*, pp. 105–8, 162–81, and 206–8.
[119] Ibid., p. 162.
[120] Bose, *Abyakto*, p. 147.
[121] Roy Bhattacharya, *Acharya*, p. 10.

dominated Bose's worldview: coping with one's violence by renouncing violence, by aggressively protecting one's likely victims.

One suspects that this concern too—along with the more general values of non-violence, the unity of life, and fear of death—overlay Bose's ambivalence towards his mother, the final paradigm of a *prakriti* that held the keys to survival and mortality. The ambivalence was the frame within which Bose fought a life-long battle with his aggressiveness and made reparations to an 'apparently sly, ruthlessly cruel' *prakriti*,[122] the ultimate public target of his conflicting passions.

Studies of socialization in India have traced to the fantasy of an inconsistent, aggressive mother the Indian's modal conflicts centring round aggression and its control.[123] Whatever the final status of such studies, the consistency between Bose's inner image of motherhood and the dominant myths of his culture is underscored by a theme he borrowed openly from his culture, that of immortality.[124]

All themes of immortality are themes of mortality too, and Bose's version also contained, within it, its inverse: a primitive, infantile fear of death. To him, his creativity was not only a positive affirmation of eternity and the re-discovery of life-processes, but also a passionate denial of lifelessness. His concept of personal achievement, academic advancement and national uniqueness was powered by this denial:

> It was a woman in Vedic time, who when asked to take her choice of the wealth that would be hers for the asking, inquired whether that would win for her deathlessness. What would she do with it if it did not raise her above death? This has always been the cry of the soul of India, not for addition of material bondage, but to work out through struggle her self-chosen destiny and win immortality.... There is, however, another element which finds its incarnation in matter, yet transcends its transmutation and apparent destruction: that is the burning flame born of thought which has been handed down through fleeting generations.[125]

[122] Bose, *Abyakto*, p. 133.

[123] G. M. Carstairs, *The Twice-born* (London: Hogarth Press, 1957); Hitchcock and Mintern, 'The Rajputs of Khalapur'; Leigh Mintern and W. W. Lambert, *Mothers in Six Cultures* (New York: Wiley, 1964); Sudhir Kakar, *The Inner World: Childhood and Society in India* (New Delhi: Oxford University Press, 1979). See a review of the relevant literature in Nandy and Kakar, 'Culture and Personality in India'.

[124] Bose, *Abyakto*, pp. 13–15.

[125] J. C. Bose, cited in Home, *Acharya Jagadis*, pp. 66–7.

Behind this affirmation lay a vision of the universe as a system of power relations, within which only power or *shakti* was indestructible.[126] But for survival one needed a special kind of power:

> Every moment so many lives are being crushed, as if they were specks of dust, as if they were worms. Do you fear the threatening speed of the wheel of life?... Brighten the divine perspective within you. You will find the universe alive, not a mass of dead matter. The humblest speck of dust is not wasted, the smallest force is not destroyed; life is also possibly imperishable. Mental force represents the ultimate triumph of life. It is by power of this force that this holy land survives.[127]

The clue to survival therefore was the maintenance of intellectual potency: 'The destruction of intellectual power is real death, hopelessly final and eternal.'[128]

Tagore, it is said, was bewildered by Bose's obsession with death and the after-life when he, more or less of the same age, was much less anxious about them. The poet did not see that this fear was concerned more with the beginning than with the end, and hence, that age had little to do with it. Bose's unconscious fear of his own anger against his mother and the fear of the retaliation it might provoke, could only be bound by the reparative concern with persons, ideas, and things which were 'undecaying' and 'beyond the reach of death'. As he stated towards the end of his life, the 'efflorescence of life is the supreme gift of a place and its associations.'[129] Presumably, one could return the gift only by containing one's destructive self, and thus avoiding the wrath of the mother and her surrogates—motherland and *prakriti*, and by protecting and affirming life in the non-living.

[126] Elsewhere, Bose also expresses his belief that the 'world has no beginning and no end' (*Abyakto*, p. 7), and power is what ultimately survives in it—'power is indestructible' (ibid., p. 14; also p. 123). It is this potency which he sought for himself and his people. Particularly so because 'mother nature is unwilling to bear the burden of inefficient lives' (ibid., p. 133). Therefore, 'Our only concerns should be how we can give up the whines of the weak, effeminate touchiness and unjust demands, and how we can shape our destiny using our own power as befits men' (ibid., p. 123).

[127] Bose, *Abyakto*, p. 160.

[128] Ibid., p. 129. See also pp. 15–16, where Bose relates his concept of immortality to the traditional concept of rebirth: 'every life has two aspects. One does not age and is undying; the temporal body covers this aspect. This cover of body remains behind. The undying speck of life builds new houses in every birth.'

[129] J. C. Bose, cited in Home, *Acharya Jagadis*, p. 85.

VIII

On many an occasion, I write without thinking... On some occasions thoughts arise in my mind without effort, and I am surprised. It is my past which is pouring these messages into my ears. The root of my heart is in India.[130]

When in adulthood Bose began to build his self-definition on identifications he had earlier disowned, his past got more deeply involved with the past of his people, torn between the pulls of tradition and the demands of the culture of full-blown colonialism.

The end of the nineteenth century found the urban élite of India, particularly the *bhadralok* of greater Calcutta, searching desperately for a self-esteem that would protect them from the severe threat to their world that colonialism now posed. Participating in this search, Bose felt the need for 'equipping Bengalis with an ideal which would make the dying race confident of its own power.'[131] But for him, given his times, the confidence had to come in a special form. Sure that everyone wanted 'to see Mother Bengal on a high throne',[132] he heard in the plaudits he received from the literati of Calcutta the call of his mother: 'In the encouragement given by you all I hear my mother's voice. Behind all of you I can always see a poor, humbly clad figure. With you, I take shelter in her lap.'[133]

Bose's participation in the struggle for a collective self-image that would counter personal feelings of inferiority—and for an individual self-image that would be more than privately valid—was perhaps bound to reflect the psychological bulwarks he had built against his deeper-lying fantasies of maternal neglect and violence. Bose was perfectly willing to read his success in science as an index of the love for him of his motherland, a 'fiery mother who ruthlessly threw her children into the cruel workshop of life' and accepted them back 'only when they won in battle fame, courage and manliness.'[134] He felt he had met her standards of excellence and had earned his share of love from a mother whose ruthlessness

[130] J. C. Bose, cited in Gupta, *Jagadish Chandra*, p. 64.

[131] Discussion with D. L. Roy, reported in Roy and Bhattacharya, *Acharya*, p. 237.

[132] Bose, *Abyakto*, p. 136.

[133] Letter to R. N. Tagore, quoted in Roy and Bhattacharya, *Acharya*, pp. 233–4; also see pp. 56, 223.

[134] Bose, *Abyakto*, p. 21.

he could rationalize as essentially the protectiveness of a loving but firm mother.[135]

> After spending a long time in foreign lands, I have come back to the lap of the loving mother, drawing my courage from the hope that she has accepted my *puja*. O mother, your blessings have secured for me recognition as a servant of Bengal and India.[136]

This was not merely a way of speaking. From London Bose had written to Tagore of a 'strange unscientific event', a striking vision he had had. The vision, Bose felt, had strengthened his love for his motherland:

> All of a sudden I saw a shadowy figure, wearing the dress of a widow; I could see only one side of the face. That very sickly, very unhappy woman's shadow said, 'I have come to accept'; then, within a moment, the whole thing disappeared.[137]

Understandably, every other acceptance became to him, ultimately, secondary.

It is not clear why the search for self-esteem and parity attained salience among the Indian élite at exactly that point of time—why, for instance, Bose felt prompted to say: 'Now the time has arrived; we must spend our entire energy in glorifying our motherland.'[138] What we do know is that the scientist worked in an atmosphere in which the early Indian intellectual and élite hopes of changing Indian society through the intervention of alien ideas and instruments of power was giving way to a greater consciousness of Indian exclusiveness, Indian categories of social change, and Indian ways to political and cultural autonomy. The climate of *swadeshi*, which included all-round attempts to revalue indigenous ideas and products and which contextualized Bose's middle years, had existed well before the movement that gave it dramatic substance after 1905.[139]

As is often the case with such movements, the search for the indigenous went with a growing awareness of native inadequacies.

135 Bose, *Abyakto*, pp. 133–9.

136 Ibid., p. 122; written in 1915.

137 Roy and Bhattacharya, *Acharya*, p. 234.

138 Bose, *Abyakto*, p. 229.

139 See Ashis Nandy, 'The Making and Unmaking of Political Cultures in India', in Nandy, *At the Edge of Psychology*, pp. 47–69.

For Bose at least this awareness was nothing new. He had already said:

> *O abhimanini* woman.... What is his status in this world on whose glory your own glory is built?... Will he, on whom you depend, be able to save you from terrible humiliations in these bad days?... Who will strengthen his arms, keep the courage of his heart indomitable and make him fearless of death?[140]

These doubts about his compatriots were intertwined with self-doubts. 'Some day surely India will see better days, but one fails to keep this in mind constantly. Imprint on my mind that this is true. I lose my power without hope.'[141]

These intense feelings of inadequacy and the parallel search for parity were partly brought about by the widening breach between the British and their subjects. The former, impressed by the new scientific discoveries and the fast pace of industrialization in the West, had not only started perceiving Indians and particularly Bengali *babus* as an essentially inferior breed, but had begun to communicate this perception to the latter.[142] As a result the goodwill between the rulers and important sections of the Bengali élite, and the mutual accommodation that had characterized their relationship throughout much of the previous century, had begun to break down.

At the same time, the internalization of values promoted by imported institutions had begun to change the impersonal perception of Indo-British differences to more personalized feelings of Indian inferiority. Working within an increasingly westernized frame of values, as against a merely westernized form of occupation or education, many Indians found themselves splintered within. They could neither live in peace with their traditions nor disown their new-found western values.

Earlier generations of Indians, sure of their traditions and of their traditional selves, rarely sought to demonstrate the superiority of native practices or ideas over foreign ones. When they opposed westernization, they often invoked the concept of an 'equal but different' Indianness. Even the early modernists who rejected

[140] Bose, *Abyakto*, p. 141. The word *abhimanini* is difficult to translate. In Bengali it roughly means a woman who is angry and hurt, but retains her affection towards the cause of her anger.

[141] Letter to Tagore, quoted in Roy and Bhattacharya, *Acharya*, p. 233.

[142] Rudolph and Rudolph, *The Modernity of Tradition*; also R. C. Majumdar, *Glimpses of Bengal in the Nineteenth Century* (Calcutta: Firma K. L. Mukhopadhyay, 1960).

aspects of Indian culture and supported all-out westernization, were a self-confident lot. Men like Rammohun Roy were not afflicted by a sense of cultural or racial inferiority. It was the internalization of modern values and the growing gap between self-perception and new social ideals that sabotaged the nineteenth-century sense of competence of Indians *vis-à-vis* the West.

However, the defeat of psychological forces is rarely total. The strategy of identifying with the West by abrogating one's Indianness as a negative identification, remained an alternative to the total rejection of the West. Such abrogation of one's cultural self now required a greater psychological effort—because of snowballing Indian self-doubts and British superciliousness—but every Indian had it before him as a latent vector. The cultural conflict between the old and the new caught up with Bose through this inner strain. If exclusivism was not unknown to Bamasundari's son, Bhagwanchandra's catholicity was for him a vital experience. If struggling for parity was the cultural 'inheritance' of the backyard man from East Bengal, Brahmoism's self-confident synthetism had also been for him a crucial exposure. It would, however, seem that the reactive self-affirmation of Bose's class and the self-doubts of western intellectuals at the time (which had been communicated to Indian intellectuals, though not to British Blimps in India) combined with the changes in his personality to encourage the exclusivist response to gradually emerge as the dominant tone in his intellectual life. 'Who could be so base as to be untrue to his salt and the soil that nourished him?' Bose was to ask, particularly when that maternal presence had given meaning to his adult attainments and struggles.[143] When the choice was forced on him, Bamasundari's son found his intellectual ideals better represented by those who spoke the language of his later self than by the cultural amalgam his father exemplified.

However, in the process of adjusting his science to the climate of Indian politics and public opinion, Bose rendered two important services to western science in India.

When Bose reached adulthood, the nativist response to the West had split the intellectual culture of modern India into two. A split between traditional and modern medical systems had already

[143] Bose, 1934, cited in Home, *Acharya Jagadis*, p. 85. Talking of his journeys to the West, Bose once said: 'Nobody there waited for me with a garland of victory, rather my powerful opponents were present in a group to demonstrate my flaws. I was all alone; the only invisible help to me was the goddess of India's fate.' Ibid., p. 148.

taken place. The native system enjoying the highest status, *ayurveda*, no longer co-existed peacefully with modern medicine (though homeopathy, with its western roots but non-modern style, in practice continued to mediate between the two). Others were talking of a Hindu chemistry and an Indian mathematics that would be separate from and superior to their western counterparts. Tagore and Aurobindo had already founded institutions that imparted 'national' education, though it was conceived in universalist terms. Vivekananda had founded a Hindu church. Some politicians, too, had begun to talk of an Indian concept of nationalism and an Indian methodology of political action.

In this environment, Bose helped arrest to some extent the compartmentalization of the sciences. His was not a technology transfer or import-substitution model, for he refused to give up Indian claims to a different form of universalism. By legitimizing his work in terms of traditions and then establishing its credentials—and, as some saw it, supremacy—in the western academic world, he obviated the need for an 'Indian' natural science for Indians, as a remedy for their gnawing feelings of inferiority. India, he was fond of saying, had a contribution to make to world science, not to a special Indian science.[144] This was another way of saying that India had an alternative world science to offer, not a separate Indian science.

Thus, by demonstrating the convergence of Indian thought and the values of modern science, years before such efforts came into vogue, Bose made possible a new commitment to universal science. He believed, rightly or wrongly, that both the content and the context of science should converge. We have already discussed Bose's search for a convergence in content. The search for convergence in context is more difficult to document. But one example could be the themes of achievement and competition which run through many of his writings. Once again, he borrowed these themes from the sacred texts of India. To a majority of Indians, the achievement criteria used by western science education and organized scientific research must have seemed an arbitrary imposition. The criteria could not but seem incompatible with learning in a society where learning seemed linked to ascription, hierarchy, and self-realization. Here, too, Bose tried to build a bridge between organized world science and his society's surviving concepts of scholarship. He extended to science the model of the Indian

[144] Bose, 1917, cited in Home, *Acharya Jagadis*, p. 66.

modernizers who vaguely sensed that many aspects of traditional high culture could come in handy in the changing world, particularly in the nascent urban-industrial, colonial culture in British India.

In other words, like other nineteenth-century social reformers in general, Bose too sought to counterpoise in Indian society the older values of organized knowledge against new values borrowed from the West which were justified by more isolated traditions in well-defined, delimited areas of Indian life where these new values did not seem that new.[145] Consequently, if to his contemporaries Bose looked a representative individual trying out the role of westernized scientist within an indigenous cultural frame, his science demonstrated, to the satisfaction of most, that the trial had been successful.

Yet it is one of the lesser paradoxes of social change that a society's success does not always coincide with an individual's. First of all, Brahmo bi-culturalism carried for Bose a load of anxiety, due to his own and his father's role confusions and failures within the compass of the modern sector and the culturally more self-assertive political climate of his adulthood. He was therefore pushed towards a form of nationalism with the powerful symbols of the Bengali *shakti* cult, especially its concept of maternal authority, to create new political solidarities. This nationalism was represented not only by the Swadeshi movement, but also by movements led by Brahmabandhav Upadhyay, Vivekananda and Aurobindo. It is no accident that the slogan *Bande Mataram* or 'Praise the Mother' became the motto of the Bose Institute, in addition to being the war cry of Indian nationalism.[146]

To the symbolisms associated with this nationalism Bose was exposed overtly in adulthood, but covertly, years earlier, by his mother. Its magical, anthropomorphic overtones and its evocation of cosmic maternal principles had a special meaning for him:

> Can a son imagine a distinction between motherland and mother? The sound of the names of mother and motherland has emerged

[145] 'We must build a western society with Indians' as even Vivekananda was to describe the task. How far they were correct in formulating the society's problems in this manner is, however, a different matter.

[146] Roy and Bhattacharya, *Acharya*, pp. 231–47; Gupta, *Jagadish Chandra*, pp. 204–6, 75–8, 84–92; also see Haridas Mukherji and Uma Mukherji, *India's Fight for Freedom or the Swadeshi Movement* (Calcutta: Firma K. L. Mukhopadhyay, 1958), esp. pp. 174–234.

spontaneously from (the) heart and spread all over India. This is because that sound has touched the inner heart of India.[147]

Thus, while he scrupulously avoided conflicts with the British power and remained friendly with the more westernized section of the Bengali élite, his creative work drew upon his early identification with a feminine principle capable of dealing magically and ritually with reality and fate on the one hand, and upon his anthropomorphic and often-magical conceptualization of personal work and personal goddesses—mother, motherland, *shakti* and *prakriti*—on the other.[148]

IX

These bits, drawn from a larger historical mosaic, mark out one possible strategy available to an Indian scientist of an earlier generation to forge a workable lifestyle, as an Indian and as a scientist. Yet this was something more than a private strategy. The 'objective' scientific imagination that linked personality, culture, and the history of science in the case of Bose was part of a larger process of social adaptation. It was within that process that cultural particulars in India could find a particular scientist, and vice versa.

India, science, and Indian science—it could be claimed that Bose was needed by all. Others might wish to reverse the relationship. Bose, they may say, needed a large canvas on which to write the history of his personal crises. I hope this narrative has made it clear that such a bifurcation of perspectives is not necessary, for between the man and the canvas—and the first and the second perspective—there was a psychologically and historically determined relationship. I shall now sketch the outlines of one possible explanation of this relationship.

With the inroads made by the western economy and education into India, the confirmations that were available to older systems of socialization and personality were lost to many. The new structures rewarded skills that had a low cultural value and negated older

[147] Letter to Subhas Chandra Bose, 1937, quoted in Roy and Bhattacharya, *Acharya*, pp. 237–8.

[148] Here lies an explanation of why many Indians found Bose to be more nationalist than his participation in politics would suggest. He shared the basic symbols, values, and adaptive strategies of Indian nationalism, and the logic of his work was perfectly compatible with the logic imputed to it by a majority of those who knew something about him and his work: the world he had created made eminent sense to his compatriots.

concepts of legitimate accomplishment. The conflicts that western science generated were part of this wider confrontation. Those who accepted this new science found it advantageous, but not meaningful. When it did not seem to disrupt what was good and moral, it looked threateningly amoral and irrelevant to deep-seated concepts of goodness and rightness. Yet this new science seemed to work and was profitable—a recognition that must only have heightened anxiety by baring one's weakening faith in tradition and the temptation of joining the ranks of the deserters.

This chasm between profitable work preference and successful work identification—between acceptance and commitment—threw up into salience individuals who could help sections of the community to integrate the new science within their lifestyle as a valid medium of self-expression. The individuals, if they were creative enough, used the apparent neutrality of the new science as a screen on which to project their loves and hates. When these loves and hates turned out to be generally held, they made the new science legitimate and the new scientists eponymous.

The changing political economy of India underwrote the process—radical permutations in family and child-rearing practices took place (often brought about by the nuclearization and geographical mobility of many families) and in adolescent socialization (brought about by westernized schools and colleges). Deviant individuals, carrying the impact of deviant socialization within them (including adult experiences that updated the impact), had now to find a medium of self-expression—a new mirror through which to see their own faces—in their professional life. We have seen how the content and form of Bose's research mirrored his early object relations, adult interpersonal skills, and the defensive strategies available to his self. Even his ability to combine good relations with the colonial authorities and cultural nationalism could be traced to the specifics of his own experiences, of life; and so could his later decline as a scientist.

In this case, as in others, the social forces that sought new carriers were the ones that produced the deviations in patterns of socialization. Here lay the basic significance of Bose's life. In trying to come to terms with himself, he produced for his community a possible new link between social needs and personal desires. At the same time, by rediscovering traditions through science, he helped retain a core of self-esteem in a people threatened by the patent supremacy and power of a foreign system; and by culturally domesticating a hostile science, he made it possible for a growing number of Indians to take to it. And had he not reached a dead end himself, he could be said to have offered the Indian scientist an enduring identity.

The road to this dead end was paved with two aspects of Bose's scientific self. First, he conceived of science as a means of experimentally demonstrating the truth of self-evident, axiomatic, general laws of nature enunciated in the sacred texts of India. To Bamasundari's son his revolutionary inquiries into *prakriti* were basically a revelatory experience. And like all revelations they included a principle of closure.

The hypothetico-deductive method in scientific creativity is so widely accepted today that this might at first seem absurd. Was it not true that even Newton's experiments were 'a means, not of discovery, but always of verifying what he already knew'?[149] The answer to this question, on the basis of Bose's life, will have to be different.

Creativity involves not merely the ability to use one's personal fantasies and the myths of one's culture; it is the ability to do so without being rigidly defensive and retaining a certain cognitive and emotional flexibility. Bose's obsessive-compulsive defences did not allow him to make full use of his own fantasy life and the mythic structure of his culture. He was forced by the demands of his personality to take his concept of Indian uniqueness beyond the culture and philosophy of science, into specific cut-and-dried theories of science and into actual research. No inspiration can carry that enormous load. (Perhaps a comparison with Newton would not be misplaced after all; the Renaissance man might have set an altogether different standard of creativity, but the difference between the young and the middle-aged Newton *is* comparable with the difference between the young and the middle-aged Bose.)[150]

Second, Bose's highly personalized science, combined with his later authoritarianism and promotion to the status of an *acharya*, did not allow him to develop an organizational base. So, he could not be fed back the experiences of his talented but professionally neutralized Indian disciples. The Bose Institute became a cult phenomenon in its founder's lifetime, and the patriarchy within it did not allow the growth of a community that could help his reality-testing. Neither the personalized style nor the paternalism was, by itself, anti-science. After all, there had been creative

[149] J. M. Keynes, quoted in J. E. Littlewood, *A Mathematician's Miscellany* (London: Methuen, 1957), p. 94.

[150] Ibid.; Manuel, 'Newton as Autocrat'.

scientists before the growth of impersonal organizations and the spread of democratic values. What compromised Bose's creativity was the combination of his personality and private theory of science, the colonial situation and the culture of the Bose Institute.

These two features of Bose's science had an especial meaning for the West. Both the revelatory mode and the personalized style had once been part of the western scientific tradition, but the West had been seduced away from them by the attractive promises of positivism and empiricism. The unsure gait of these philosophies of science and the conflicts they generated in Bose's time—such as the anxieties they aroused in many western intellectuals because of the violence increasingly associated with science and technology—made a section of the western élite look back guiltily towards its own past, as reflected in the scientist-savant from the East. In his worldview, the recovered past of a modern society and the cultural compulsions of a traditional society seemed to intersect. But it was also a worldview through which Bose's lay disciples in the East and the West exercised immense pressure on him to produce a science that would contain their own anxieties *vis-à-vis* modern science. Bose did not have the personality resources to resist these pressures. He went on widening his reference group to make sense simultaneously to scientists and to the larger public in the East and the West. Ultimately, what had been a strength in his struggle to achieve compatible scientific and cultural identities, became a charter for an unconditional surrender to the public demands on his science. Put simply, Bose became inauthentic as a critic of western science, too, and fell a victim to his public image.

There is another meaning behind Bose's movement from his 'clever' early works to the deeper successes and failures of his later phases. All his life he had searched for a personal identity that would be legitimate by being the prototype of an Indian scientific identity. As his early personality conflicts got resolved into a relatively stable inner man, Bose owned up parts of himself he had formerly strenuously rejected. This rejected self, with its measures against fears of one's own inner violence, its over-compensations for feelings of neglect and deprivation of nurture, its defensive orderliness and authoritarianism, was more compatible with his later science. It was at this plane that some themes in the contemporary culture of science—the stress on achievement, competition and organizational skills, hardheaded positivism, and belief in science as a value-neutral instrument of power and as a means of subjugating nature—contradicted a part of the substance of which he was made. For many years they had found a haven in the residuals of his apparently idyllic relationship with his father. But, while struggling for a personality integration where his first and deepest identification could find a place, Bose had to move from an Indian version of world science to an Indian—and paradoxically western—version of Indian science.

What then is Bose's lasting relevance to science and to India? One possible answer is that while Bose sought and found a sanction for a private experience which was recognized at that time as 'scientific creativity', he paid back his debt to a culture which had tangibly intervened in his personal science—supporting, sanctioning, and articulating it with his own needs and history. Bose's scientific universalism had to include a concept of what world science could and should do for India.

This brings us back to the central problem which Indian science faced in Bose: how to reconcile the Bose of spatial and temporal meanings with Bose the space-and-time defying scientist? It is obvious that whatever evidence of such a reconciliation one finds in Bose was only partly shaped by the formal structures of western science and Indian thought. Plant physiology, it is true, was stagnant when Bose entered the field.[151] And the diffusion of the dominant paradigms of the discipline allowed deviant scientists like Bose more elbow room and a larger audience than they would otherwise have got. It is also true that some of the ideas Bose used were straight out of his reading of the classical texts of India. But what gave him his appeal was that he could reconcile, even if temporarily, the text and the culture of modern science with the shared experiences that had been banished from consciousness by the scientists or disowned by the societies of the East and the West.

Scientific creativity, like any other form of human creativity, assumes the ability to use one's less accessible self in such a way that the primordial becomes meaningful to the community and the individual scientist. Out of this ability comes not only the creative scientist's sense of being driven, but also his distinctive approach to concepts, relationships, and operations, the order that he imposes on his data, and the limits he sets on his insights. The scientific community prescribes where and when professional assessments

[151] D. M. Bose, 'Jagadis Chandra Bose'.

begin, but it can never fully control what at any point of time is accepted as objective, impersonal, and formal scientific knowledge.

The scientific community can do worse. It can begin to collaborate with the larger society to discourage the self-examination that may accompany or follow from a creative effort. Any such examination is an inquiry into the psychological foundations of science and society and, like individuals, both have built-in resistance to self-inquiries. Science tries to build self-perpetuating paradigms and societies try to build self-validating solidarities. Both have a vested interest in establishing a link with the fears, hates, hopes, and ideals of individual scientists. Neither can afford to recognize that this search for significant solidarities can degenerate into a subtle pressure on individuals and groups to settle for palliatives which, instead of widening their awareness of their outer and inner worlds, would only sanction and sanctify their existing desires and anti-desires.

It was in this process of consolidation that Bose was so cruelly caught. It is true that his culture and the state of science *had* shaped a part of his creativity and it was an important part. But it is also true that when in the end he failed science, in a deeper sense his society and the scientific estate had failed him too.

Part Three

The Other Science of Srinivasa Ramanujan

An Essay on the Public and Private Cultures of Knowledge

Towards the beginning of the twentieth century, the culture of science in India saw an all-round attempt to Indianize modern science. At one level it was an attempt to cope with the contradiction between the older concept of science, embedded in the worldview of an old civilization, and the demands of a newly dominant, alien culture of science. But at another level it was an attempt to cope with a more persistent contradiction within modern science itself. While a scientific estate survives on its faith in the universality of science, supported by a belief in the organized, disinterested, sceptical objectivity of scientists as a group,[1] a central and necessary feature of the social system of science is the scientist's faith in dominant disciplinary dogmas, euphemized in recent times as paradigms.[2] Evidently, the openness of mind expressed in universalism and rational scepticism, and the closedness nursed by scientific socialization and expressed in paradigmatic faith, contradict each other and demand psychological functioning simultaneously at two levels.[3] Evidently, too, for

[1] R. Merton, 'Science and Democratic Social Structure', in *Social Theory and Social Structure* (New York: Free Press, 1957), pp. 550–61.

[2] The best-known examples are Michael Polanyi, *Personal Knowledge* (Chicago: University of Chicago Press, 1958), and *Science, Faith and Society* (Chicago: University of Chicago Press, 1964); T. S. Kuhn, *The Structure of Scientific Revolution* (Chicago: University of Chicago Press, 1962); see also the interesting brief statement by M. Mulkay, 'Cultural Growth in Science', in B. Barnes (ed.), *The Sociology of Science* (Harmondsworth: Penguin, 1972), pp. 126–42.

[3] As is well known, Kuhn in *The Structure of Scientific Revolution* has seen this contradiction as a cultural tension within the estate of science. I speak here of the extensions of this tension within the personality of the scientist.

many Indian scientists of the time this contradiction took a particularly painful form. While western science seemed congruent with their newly internalized concept of a universal science, they had to reconcile this concept with their deeper selves and inherited concepts of science and the scientific method and, through them, with their own society's distinctive traditions and entrenched symbols.[4] It is the problems of individual creativity and professional identity posed by these contradictions in one scientist caught in the hinges of historical change that I shall discuss here.

The scientist I summon as my witness is Srinivasa Ramanujan (1887–1920), a favourite textbook example of natural genius in mathematics. I shall try to show how the two contradictions in the scientific culture became in him a part of creative tension and defined important aspects of his scientific self. In the process, I hope to identify a prototypical style of reconciling the relatively universal and rational structure of outer science and the extra-logical, culturally and psychologically bound, inner science in India.[5] This style, apparently non-rational and anti-science, has taken the culture's inner science close to the core of the objective, secular, and universal body of the outer science but—this is its strength—without disturbing the authenticity of either.

Two doubts may persist about the relevance of such an exercise. The first of these is roughly the position of Ramanujan's closest academic friends, G. H. Hardy and J. E. Littlewood.[6] If I understand them correctly, their argument is that Ramanujan's origin in a distinctive culture and time, his religious faith, worldview, and personal motives had little bearing upon his mathematics. To these friends, Ramanujan did not seem to have

[4] On access to preconscious fantasies as a condition of creativity, and on regression at the service of the ego as a source of creativity, the pioneering works were by L. Kubie, *Neurotic Distortion of the Creative Processes* (Kansas: University of Kansas, 1952); and E. Kris, *Psychoanalytic Explorations in Art* (London: Allen and Unwin, 1953).

[5] Obviously these concepts of outer and inner sciences are related to Einstein's distinction between science as 'something existing and complete'—as the 'most objective thing known to man'—and 'science in the making, science as an end to be pursued', which is 'subjective and psychologically conditioned'. See his 'Address at Columbia University, New York, January 15', *Essays in Science* (New York: Wisdom Library, 1934), pp. 112–14, esp. p. 112. See also P. P. Wiener and A. Noland, 'Roots of Scientific Thought; A Cultural Perspective', in *Roots of Scientific Thought* (New York: Basic, 1957), p. 3.

[6] Hardy held a somewhat different view at the beginning. See G. H. Hardy, 'Notice', in G. H. Hardy, P. V. Seshu Aiyar and B. M. Wilson (eds.), *Collected Papers of Srinivasa Ramanujan* (New York: Chelsea, 1962), pp. i–xxxvi.

any serious interest outside mathematics, and his mysticism and religion 'except in a strictly material sense played no important part in his life.'[7] He was, in fact, according to one of them, an 'agnostic in its strict sense', and his conformity to his cultural traditions was determined not by any faith or commitment, but by his acute this-worldly pragmatism. From this point of view, Ramanujan's professional self could be parsimoniously explained in terms of his sane and shrewd rationality, his lack of a proper mathematical education, his killing poverty, and his 'natural' genius.

Underlying this position, I suspect, is a feeling that 'Ramanujan's mathematics was outshone by the romance of his mathematical career' and a tinge of sorrow that Ramanujan did not turn out to be a greater mathematician and a plainer man.[8] To these friends, what mattered ultimately was the content of his mathematics, the clue to which would have to be sought in the autonomous, intellectual structure of his work, the drama of his life being at worst a red herring and at best an irrelevance.

Against this we can place the observations not only of the starry-eyed Madras mathematicians and Indian friends of Ramanujan, but also of some rather hard-headed scientists who knew him well. E. H. Neville says in Ramanujan's obituary: 'He had serious interests outside mathematics; he was always ready to discuss whatever in philosophy and politics had last caught his fancy.'[9] And Prashanta Mahalanobis was convinced that, at one stage of his life at least, Ramanujan was striving towards a Vedantic mathematics and would have been happier to see his philosophy vindicated by his mathematics than in being recognized as a mere mathematician.[10] One's own view of one's work may be misleading but even B. M. Wilson, who spent some of his most creative years on Ramanujan's work, feels that the latter 'remained to the end an unorthodox mathematician. Almost everything he published bears, either in matter or the treatment, the imprint of his personality; his was highly individual work.'[11] Thus, it may not perhaps be considered

[7] G. H. Hardy, 'Introduction', *Ramanujan—Twelve Lectures Suggested by His Life and Work* (Cambridge: Cambridge University Press, 1940), pp. 4–5.

[8] J. E. Littlewood, *A Mathematician's Miscellany* (London: Methuen, 1957); and Hardy, 'Introduction'.

[9] E. H. Neville, 'The Late Srinivasa Ramanujan', *Nature*, 1920, *106*, pp. 661–62.

[10] See Section III below.

[11] B. M. Wilson, Unpublished MS on Ramanujan, Add. M.S.G. 107C, Trinity College Library, Cambridge University; hereafter Trinity Papers. Even Hardy in *Ramanujan* recognizes that the Indian was, at one level, consistently 'odd and

an instance of uncompromising psychologism if in the following narrative I seek clues to some aspects of Ramanujan's creativity and professional identity in his individuality and personal history.

The objection of Hardy and Littlewood can be met in another way. Those interested in science as a closed system of formal incremental knowledge may not perhaps find it impossible to write a history of world mathematics without mentioning Ramanujan though it will be very difficult. But to those interested in the nature of scientific creativity, it is not merely what Ramanujan added to the formal structure of mathematics which determines his importance in the history of science; it is rather the unique internal equipment he employed as an essentially self-taught genius and the opportunity he gives one to study creativity as a psycho-social process by diverting attention from the net additions to knowledge made by individuals. Like J. B. Priestley's time plays, Ramanujan's life allows one a glimpse into the alternative world of 'what could have been' without the interventions of the dominant culture of science and the dominant mode of scientific socialization.

The second objection to a psychological study of Ramanujan's life is actually a derivative of the first, but it is formulated in terms of Ramanujan's discipline rather than personality. It says in effect that pure mathematics is itself so peculiarly abstract and so far removed from the real world of events, feelings, and objects, that it is perhaps the only 'true' science with no determinants outside its own world. There is obviously some truth in this argument, but it is very partial. The pursuit of a pure science like mathematics demands not only certain cognitive skills but also an emotional apparatus that will allow, enjoy, and facilitate certain ways of problem-solving. To say that pure mathematics is, unlike most other human pursuits, culture-free and unencumbered by human emotions is to say that only some kinds of persons can become creative mathematicians.

In addition, those who stress the autonomy of the world of

---

individual'. See also Hardy's 'Notice', where he makes the point that Ramanujan 'had a passion for what was unexpected, strange and odd' and that 'all his results, new or old, right or wrong, had been arrived at by a process of mingled argument, intuition and induction, of which he was entirely unable to give any coherent account'. See also L. J. Mordell, quoted in Suresh Ram, *Srinivasa Ramanujan* (New Delhi: National Book Trust, 1972), pp. 46–7.

mathematics in connection with Ramanujan's life—predictably, they are mostly mathematicians—seem to miss an important point. Ramanujan depended very little on the existing structure of mathematics, and his work can hardly be called the logical development of the formal mathematical knowledge of his time. His exposure to modern mathematics was next to nothing, and his method was predominantly 'induction from particular cases in the crudest sense'.[12] He was an artist who was satisfied if he could convince himself.[13] Littlewood is more explicit: Ramanujan, he says,

> . . . had no strict logical justification for his operations. He was not interested in rigour. . . . If a significant piece of reasoning occurred somewhere, and the mixture of evidence and intuition gave him certainty, he looked no further.[14]

Whether mathematicians like it or not, the products of such an approach have to draw upon the inner resources of the creative scientist himself and depend on his inner checks and reality testing, rather than on a standard reference group of other scientists. Anyone interested in the work of such a self-sufficient genius simply cannot avoid considering these inner resources and checks.

Also, just as Ramanujan's life tells us about the process of creativity by presenting 'alternative scenarios', his successes and failures tell us not merely about the structural growth of mathematics but also the cultural changes in science—the process through which science moves from one culture of creativity to another. At a time when science itself is examining the social and psychological factors that have led it to value its cumulative objective mass of knowledge more than its non-cumulative, non-rational, time and space-bound culture (and in the process lose its soul), a study of Ramanujan gives one the chance to speculate about a future science which might allow one to integrate the speculative, normative, and aesthetic factors with the logical, rational, and empirical ones. It permits one, at another level, to examine the living presence of the scientist within his science and the shifting relationship between observer and observed, not as something merely incidental, but as part of the essential core of science.

[12] Wilson, 'Trinity Papers'. See also note 11 above.
[13] J. E. Littlewood, Letter to Hardy, Add. MS a 94 (1–6), Trinity Papers.
[14] Littlewood, *Mathematician's Miscellany*, p. 88.

To the extent that the text and culture of modern science are moving towards greater recognition of these epistemic and methodological problems, the personal identity of Ramanujan is as relevant to science as it is to the evolving discipline of the science of science. To take a narrower view, it may also yield concepts in terms of which one could study the life histories of a number of Indian scientists of the last hundred years as records of an unconscious and daring—though mostly crude and inelegant—effort to move towards such a new science. It is as if a major civilization, in trying to meet its own problems of survival, was trying prematurely through these men to anticipate if not tackle some of the fundamental issues in the philosophy of modern science.

This point could be made in yet another way. Each phase of science has, as its correlate, a distinctive philosophy of science embodied in a congruent world image and a unique inner concept of science which gives scientific activity its subjective logic. When the Enlightenment in Europe produced a new concept of science in its attempts to give a prominent place to empircism and experimentation in scientific activity, it reduced the legitimacy of speculative and rationalist elements in the modern culture of science. The delegitimization was then necessary for the growth of modern science. But while some permanency may attach to fundamental scientific problems, none attaches to their solutions or to the contexts in which these solutions are offered. Thus, as the environment of science has continued to change, many recessive elements in the culture of science, once rejected as anti-science, have again come to the fore in response to the changed concerns of the scientific estate.

Here lies the major significance of Ramanujan in the history of world science. His professional life gives us a rare opportunity to study the survival and dissolution of cultures in the minds of scientists, and beautifully illustrates the attempt to produce a valid science with an 'anachronistic' concept of science and with the psychological and cultural equipment reportedly more suited to an earlier phase in the growth of science. Whether his specific orientation and skills can become relevant again to world science is, of course, an open question which future philosophers and historians of science may be able to answer better.

My knowledge of mathematics being what it is, I am sure that the following narrative does not explain Ramanujan's mathematics and cannot therefore deal with some of the issues raised above with even the barest sophistication.[15] But I hope that it explains something of the mathematician: the distinctive meaning his creations might have had for him and his group; the way in which his time and his society lived within him, sometimes validating and sometimes threatening his professional self; the context of outer processes and inner needs that he used in defining his own scientific identity; his attempts to solve through his science the contradictions between a long tradition of indigenous science and the felt need to participate in world science on a new footing; and finally, between the demands of the world of formal mathematics and the history of his own selfhood.

II

Srinivasa Ramanujan was born in a poor Tamil Iyengar family in 1887 in Erode in Tamilnadu, not very far from the city of Madras. Following local custom, his mother had gone to her father's home for the birth of her first child. Ramanujan was his first name. Following South Indian custom, he always wrote it second.

Only a substantial amount of piety accumulated over previous lives ensured one's birth as a Brahmin. It was even more of an achievement to obtain an Iyengarhood. Iyengars traditionally enjoyed the highest status among Tamil Brahmins, in a region where Brahminism itself, in Ramanujan's time at least, was culturally more hegemonic than in most other parts of India. In addition, the gap between Brahminic and non-Brahminic cultures was one of the widest in Tamil society. This combination of hegemony and exclusivism perhaps bound part of the anxiety associated with poverty which one expects in a family of this type in other societies. Brahminism might also have given a special meaning to the poverty by linking it to austerity. All said, Ramanujan's ritual status gave him one message, his poverty another. And he might have been forced early to learn to use the Brahminic worldview to rationalize and compensate for his poverty.

Like most gifted scientists, Ramanujan was the eldest son.[16] His

[15] Fortunately, however, excellent reviews of his mathematics and his intellectual career are available. The best known of them is naturally Hardy's *Ramanujan*.

[16] Cf. F. Galton, *English Men of Science* (London: 1874); J. M. Cattell and D. R. Brimball, *American Men of Science* (Garrison: Science, 1921); Anne Roe, 'A Psychological Study of Eminent Psychologists and Anthropologists, and a Comparison with Biological and Physical Scientists', *Psychological Monographs*, 1953, 67 (2), Whole Number; and *The Making of a Scientist* (New York: Dodd Mead, 1952).

two younger brothers were born several years after him and thus he also had the upbringing of an only son. Such experiences have a special meaning in many cultures. They are even more meaningful in a culture that is partial to sons and considers them props to the family economy, as old-age insurance, and sources of comfort in life after death because of their ritual role. Moreover, Ramanujan was born to his parents after years of childlessness. Almost by definition, one may say, he was a subject of heightened care, pride, and parental expectation.[17]

After Ramanujan's birth his mother returned to her husband's family—consisting of her husband and parents-in-law—at Kumbakonam, a small, rather typical south Indian town, midway between Madras and Cochin. It is here that the mathematician spent most of his childhood and youth. Kumbakonam was a traditional town, enjoying some status among Tamils for the alleged subtlety and refinement of its people, especially its Brahmins. However, even their fans never failed to point out that this smoothness in social style was associated with a certain sly deviousness, interpersonal distance, and insincerity. The popular perception of the town paralleled the popular perception of Iyengars who were also supposed to have a polished manner which hid shrewd ruthlessness behind a façade of traditional refinement. Stereotypes are important not for the element of truth they rarely contain, but for the feelings they invariably communicate. From these shared images of yester-years, one gets the feeling that both Kumbakonam and the Iyengars of Kumbakonam carried the connotations of tradition, élite status, and that delectable touch of decay often associated with traditional élite status.

Ramanujan's family lived in a town and though most of its members, including his mother, were educated, they were deeply conservative. What appears more remarkable—and the traditional Indian attitudes to poverty and suffering perhaps had something to do with this—is that the family, especially Ramanujan, had a certain unselfconscious confidence that was expressed in a serene acceptance of, and pride in, their own way of life. It is true that

[17] The Indian first child's world is however also vulnerable in its own special way. While his self-regard is protected by his interpersonal environment, it is also overly dependent on his ritual role and upon his being an anchor of the forces of cultural continuity within the family. As a result, he frequently becomes a battleground for different norms of social living and different expectations about his social role. See a discussion of this dynamic in J. C. Bose above.

there was Ramanujan's ordinal position among the children; his sense of competence does resemble the typical psychological profile of the first-born.[18] But his self-confidence was also protected by his family's self-confidence. His colossal success in the modern sector was eventually to shatter that confidence. So it would appear from the whining, supplicatory letters that his younger brothers wrote to his English mentors begging for trivial favours after his death.[19]

One can only guess at the emotional consequences of this combination of poverty, partial modernization, and orthodoxy. At one level it meant a combination of enforced simple living, acquired middle-class thinking, and persistent Brahminism. At another, it gave the family a distinctive ability to live off the modern sector without being swept out of their traditional ways and—which is the same thing—an ability to cope with the demands of urban living at the margins of India's expanding tertiary sector with the psychological assets and skills of an older lifestyle.

The family was deeply devoted to the goddess Lakshmi Namagiri, Vishnu's consort in local mythology, and the goddess of worldly success and wealth. It could be that it was their poverty that induced the family to so venerate the goddess; more probably it was Namagiri's dominant status among the Kumbakonam Brahmins. Kumbakonam had a famous temple where dark granite icons of Namagiri and Narasimha were enshrined. Of the two, Namagiri was the more important deity locally; the temple was named after her, and she was considered the more powerful.

Whatever the source of this special veneration, both the two family myths that sought to explain Ramanujan's creativity to his bewildered relatives involved the goddess. The first version was that Ramanujan was born after his grandparents had fervently prayed to Namagiri for a grandson and after his grandmother had in a trance, just before her death, assured his mother that she

[18] Cf. A. Adler, *The Individual Psychology of Alfred Adler*, edited by H. L. Ansbacher and Rowena Ansbacher (New York: Basic, 1956), pp. 376–82; and Irving Harris, *The Promised Seed, A Comparative Study of Eminent First and Latter Sons* (Glencoe: Free Press, 1964).

[19] Oscar Lewis seems to suggest that the older generation, being more rooted in traditions, often do not share the sense of inadequacy of the younger ones who are no longer as secure in tradition due to their acquired modernity. Oscar Lewis, *Children of Sanchez* (New York: Random House, 1961).

would continue to speak through her grandson. According to the second, Ramanujan suddenly showed precocity after Namagiri, in one of his childhood dreams, wrote upon his tongue. An accident that must have seemed consistent with these myths was Ramanujan's noticeable physical resemblance to his mother and grandmother. His younger brother was to put it succinctly in a letter to Hardy: 'I have a corpulent mother who resembles my brother in all his physical features.'[20]

In some cultures, one could talk with some confidence about all this contributing to a diffusion of sexual identity. But in the Indian tradition, with its more complex and less differentiated distribution of gender-specific social characteristics, one can only talk cautiously of a possible heightened awareness in Ramanujan of his identification with his mother, a greater acceptance of certain feminine sensitivities and intuitive behaviour—both predictors of creativity in men—and a deeper concern with what has often been associated in many a peasant culture with primordial magical power, motherliness.[21]

Ramanujan's favourite deity, however, turned out to be Narasimha, a particularly aggressive incarnation of Vishnu, the supreme protector in the Hindu pantheon. According to popular mythology, Vishnu was once compelled to appear as Narasimha or Man-Lion to protect his earnest devotee Prahlad from the homicidal

[20] S. Lakshmi Narasimhan to G. H. Hardy, 29 April 1920, Trinity Papers. See also S. Ranganathan, *Ramanujan, the Man and the Mathematician* (Bombay: Asia, 1967), pp. 72, 126. Ranganathan's earnest and over-determined mystical explanation of Ramanujan's creativity should not blind us to the important clues that his book provides to the shared myths and fantasies in Ramanujan's environment and the meaning of his science to his community.

[21] On the likeness between Ramanujan's upbringing and early experiences of the typical homosexual in the West, see I. Bieber *et al., Homosexuality: A Psychoanalytic Study* (New York: Basic, 1962), particularly pp. 22–4, 44–53, 86–93; R. Green, *Sexual Identity: Conflict in Children and Adults* (New York: Basic, 1974), Ch. 15; and M. T. Saghir and E. Robins, *Male and Female Homosexuality* (Baltimore: Williams and Wilkins, 1973), Ch. 2. On the association between creativity and bisexuality, see for instance, Frank Barron, *Creativity and Personal Freedom* (Princeton: Van Nostrand, 1968); and *Creative Person and Creative Process* (New York: Holt, Rinehart and Winston, 1969); D. W. Mackinnon, 'The Personality Correlates of Creativity: A Study of American Architects', in P. E. Vernon (ed.), *Creativity* (Harmondsworth: Penguin, 1970), pp. 289–311. See the reviews of relevant literature in Anne Roe, 'Psychological Approaches to Creativity in Science', in M. A. Coler (ed.), *Essays on Creativity in the Sciences* (New York: New York University Press, 1963), pp. 153–82; and Frank Barron, 'The Psychological Study of Creativity', in *New Directions in Psychology* (New York: Holt, Rinehart and Winston, 1965), vol. 2, pp. 1–134.

rage of his demon-father Hiranyakashipu, and to slay the latter, tearing him to pieces. In this leonine godhead and protective male authority Ramanujan was to find later the ultimate source of his inspiration and creativity in mathematics.

One wonders why a man born as the gift of Namagiri should become a devotee of Narasimha. One possible explanation is the correlation in Tamil culture between low traditional status and an agrarian lifestyle on the one hand, and allegiance to feminine deities on the other;[22] so that ambitious first-generation city-dwellers often felt tempted to switch their devotion to more status-giving gods. But this explanation does not quite fit Ramanujan's unselfconscious lifestyle, and one is tempted to ask questions about his early developmental history which can never be answered fully. Was it, for instance, an expression of his deepseated ambivalence towards the family deity Namagiri and the maternal authority she represented? Or was it a denial of his feminine self, closely linked to the mother goddess through the family myths about his birth and genius? Was identification with Narasimha the symbolic expression of legitimate destructive urges in an Oedipal situation? Or was it a compensation for the absence of an authoritative male role model in the family? Against what source of threat was Narasimha's aggressive protection sought?

Let us start with what is proverbial: Ramanujan's closeness towards his mother Komalatammal. Even when he was alive, it was clear to all who knew him that Ramanujan's psychological life was primarily an unfolding of his complex relationship with his mother. That his Indian contemporaries and biographers should find this closeness striking is particularly significant, because the taboos that attached to the father–son relationship did not apply to mother–son intimacy in the Indian tradition. At a certain level of awareness, the mother in the Indian family was not merely expected to be the earliest prototype of an intimate warm authority, but also the first target of defiance. It was as if part of the Oedipal hostilities towards the father, conspicuously low in the Indian system, found expression in the Indian son's intimate ambivalence towards his mother.[23] If, after allowing for all this, some Indian contemporaries

[22] At least this is the impression one gets from H. Whitehead, *The Village Gods of South India* (Calcutta: Association Press, 1921).

[23] See a brief discussion of this in Dhirendra Narayan, *Hindu Character* (Bombay, Bombay University Press, 1957); Sudhir Kakar, *The Inner World: Childhood and Society*

found the mathematician's attachment to his mother noteworthy, it must have been either atypically deep or atypically conflictual or both.

A few facts are in any case well known. First, in all major decisions in Ramanujan's life, his mother had the last say, either directly or indirectly. Being a 'shrewd and cultured lady', she never made her authority too obvious or her 'voice' too strident. Rather, until Ramanujan's wife entered the household, she ruled the roost with a quiet firmness which, even if in some ways culturally typical, was nevertheless remarkably efficient.[24] Apparently it was a benevolent matriarchy which depended less on direct intervention in the son's affairs and more on indirect control over him through his inner concepts of authority, moral rectitude, and transgression.

Second, the son associated the mother with internal demands for performance. Perhaps it was this that some of my informants had in mind when they said that Ramanujan's mother made him what he was. Early in his life she established with him a long-term collaboration in the magical manipulation of mathematical symbols, numbers, and matrices for various occult purposes. Komalatammal was an astrologer and a numerologist herself and could recite her son's horoscope from memory.[25] She is said to have foreseen—which means she and her acquaintances thought that she had foreseen—Ramanujan's meteoric rise to greatness as well as his premature death.[26] By the time the son grew up, he had acquired the mother's

---

in India (New Delhi: Oxford University Press, 1979); and Ashis Nandy, 'Sati or a Nineteenth Century Tale of Women, Violence and Protest', in *At the Edge of Psychology: Essays in Politics and Culture* (New Delhi: Oxford University Press, 1980), pp. 1–31.

[24] Local myths would have it that this particular style of control was also typical of Iyengar womenfolk. Although there may or may not be any substance in this, there is less doubt about the high status of women among the Iyengars. There is even less doubt about the family tradition of this particular Iyengar family. Ramanujan's grandmother was known for her dominant style and his grandfather for his docile submissiveness.

[25] She may have also been a part-time 'witch doctor'; Ramanujan's maternal grandmother almost certainly was. See the reminiscences of M. Anantharaman in P. K. Srinivasan (ed.), *Ramanujan, 1: Letters and Reminiscences* (Madras: Muthialpeth High School, 1967), pp. 97–8.

[26] Ramanujan's biographer P. K. Srinivasan showed his horoscope to a few astrologers without telling them whose it was. All of them predicted an early death. It is possible that both Ramanujan and his mother had made the same prediction for themselves, and it became a self-destructive, self-fulfilling prophecy.

tastes. He showed his strong identification with her not merely by claiming proficiency in astrology and precognition, but by making the crucial association throughout life between femininity and the ability to handle magically the world of sacred numbers. It was as if his joint excursions with his mother into astrology and other cognate 'sciences' marked out the design of a joint intellectual venture within which all his other collaborations had to be fitted.[27] I shall have something more to say on this later.

The mother–son collaboration also included an element of aggressive competition for control and power, though only at a symbolic level. Both Ramanujan and his mother were fond of playing an indigenous board game called the fifteen-point game, where one player has fifteen pieces standing for sheep while his opponent has three pieces symbolizing three wolves. When a sheep is surrounded by wolves, it is eaten and the player loses a piece. When a wolf is surrounded by sheep it is immobilized. It is in the nature of the game that the person playing the wolves generally wins. However, young Ramanujan mathematized the game within a few days and never again lost a single game to his mother, whether in the incarnation of a sheep or a wolf.

Lastly, an accident of genetic history—or fate, as Ramanujan and many of his contemporaries would have said—endorsed his identification with his mother. Ramanujan not only resembled his mother and grandmother in looks, he had, despite his chubbiness, a delicate and conspicuously feminine build and appearance. His velvety soft palms and long tapering fingers have been frequently commented upon. Some of the femininity had less to do with congenital physical characteristics than with acquired habits; many noticed his practice of walking with his hands out, palms down, fingers outstretched and pointing laterally, a mannerism which, in India at least, would be considered rather feminine. Young Ramanujan's aversion to outdoor sports and his manifest shyness went with this body image. They strengthened his awareness of his feminine self and powered his attempts to integrate this self as a crucial element of his self-identity.

However, one must hasten to add that a certain extroverted activism is not as clear an indicator of masculinity in India as it is in the West, for, even if the social definitions of gender roles are well

[27] The depth of this relationship was recognized by everyone, so much so that after Ramanujan's death it was his mother who was paid his pension, not his wife.

delineated in Indian culture, the psychological definitions are more confluent. Under a number of specific conditions, femininity can become an indicator of a higher form of masculinity. Not that there are no St Francises of Assissi in the West, but perhaps they are possible in many more sectors of life in India.[28]

On the other hand, Ramanujan's favourite lion godhead certainly contrasted with the diffident, retiring, male authority figure that he encountered in his father. Kuppuswamy Srinivasa Iyengar was a *gumasta* or petty accounts clerk in a cloth merchant's shop, earning a paltry salary. It is tempting to believe that the father's familiarity with numbers had something to do with the son's aptitude. But, as we have seen, there were already more powerful influences working on Ramanujan. At most, the father's profession possibly deepened the impact of the mother's passion for numbers.

Kuppuswamy Srinivasa was, by all accounts, an unassuming ineffective man who rarely got involved—or, for that matter, was involved—in his son's life. By itself, this weightlessness of the father as an immediate authority was not culturally atypical,[29] especially in a linear joint family with living grand-parents where it was customary for the father not to pay much attention to his children. This was one of the means by which the society underplayed the boundaries of nuclear units within a joint family. It was the mother's personality, I suspect, which made Kuppuswamy Srinivasa seem even more inconspicuous and non-protective than he actually was.[30] She was openly contemptuous of her husband,

[28] See on cognate issues Erik H. Erikson, *Gandhi's Truth: On the Origins of Militant Nonviolence* (New York: Norton, 1969).

[29] For instance, M. S. Gore, The Impact of Industrialization and Urbanization on the Aggarwal Family of Delhi Area, Ph. D. dissertation (Columbia University, 1961), University Microfilms (Ann Arbor, Michigan), pp. 2–59; G. M. Carstairs, *The Twice Born* (London: Hogarth Press, 1957); and P. Spratt, *Hindu Culture and Personality* (Bombay: Manaktalas, 1966).

[30] In fact, when counterpoised against each other, Ramanujan's parents seem to approximate Anne Roe's description of the early family environment of social scientists, rather than of mathematicians (see Roe, *The Making of a Scientist*). On the relationship between creativity in western scientists and maternal possessiveness and hostility combined with a distant father, see also B. T. Eiduson, *Scientists: Their Psychological World* (New York: Basic Books, 1962). It would, however, be hazardous to see in this combination a source of Ramanujan's concept of science as a cross between formal thinking and mystic experience. For, in this case, the crucial datum might have been not the actual content of parental models, but the extent to which they were a deviation from cultural norms.

particularly of his intelligence and, as the reader may have suspected from this narrative, tried to find in her son a substitute for her husband.

The distance between father and son was later increased by Srinivasa Iyengar's total inability to understand his son's concerns and ambitions. One gets the feeling that the pace of Ramanujan's success overawed him. But this interpretation could be overdone, for long before Ramanujan shot into fame, Komalatammal was taking an active interest in her son's mathematics and Srinivasa was a non-person in his son's life. Komalatammal might occasionally have been a trial to her son, but she was also the only intimate figure to have genuine interest in things that were central to him. Perhaps it would not be an exaggeration to say that it was she who discovered Ramanujan.

To Srinivasa the accounts clerk, on the other hand, what mattered were Ramanujan's status as the eldest son in the family and the rupee value of his numerous ill-paid jobs, stipends, and fellowships. He was a poor man with the poor man's blend of practical sense and cynicism. In any case, he had little reason to introspect or ponder over the 'finer aspects' of his relationship with his son, given that this father–son distance was not uncommon in an Indian family.

The result of all this was that the mathematician rarely, if ever, talked about his father. His biographers, too, barely mention him. Only one, almost accidentally, records the fact that he survived his son by a few months. We know nothing about how he reacted to the boy Ramanujan or to the adult; nothing even about the way he adjusted to his son's fame, and to what was, by the family's standards, prosperity.[31]

Lastly, as already mentioned, Ramanujan's brothers were much younger and this ensured him the lonely, fantasy-rich childhood of an only son. From their stray correspondence with Hardy one gets the impression that they were simple folk who tried to capitalize on their brother's fame in their own naïve fashion. It is an indication of the low ambitions of the rest of the family that they aimed no higher than at clerical posts in government departments which they hoped to get by pestering Hardy, in the ornate Indian English of an

[31] An indicator of the modern environment in which Ramanujan's illustrious contemporary, Jagadis Chandra Bose, was brought up is that his father was an active intervening figure who was, if anything, more modern than his son. See Part Two above, esp. sections III and IV

earlier generation, for testimonials. Once they became tenured clerks, they lived out their placid lives like good Tamil Brahmins, and Ramanujan perhaps became for them, after his lonely death, one of those mythic figures in an Indian family whom one is always expected to live with but never live by.

Very little is known about Ramanujan's childhood and adolescence. Indian civilization has never made a clear distinction between the legendary and the empirical. And the former, conceptualized in terms of the primitive logic of dreams and fantasies, is frequently made to yield a hypothetical model from which the 'historical realities' of a person or a group are deduced. Such a model, I suspect, is comprehensible mainly in terms of constructs that link the group's interpretation of the person to the person's relevance for the group.

From the more concrete bits of information, one knows that Ramanujan went to school at the age of five and passed his primary school examination in 1894. He stood first in the whole district and was exempted from paying part of the tuition fees. One also knows that in common with many great mathematicians, he showed early signs of mathematical talent;[32] and that he found most of his mathematics teachers more ignorant than himself. This experience too, judging by Hardy's impression, is quite a common one for creative mathematicians.[33] However, in his relationship with these teachers he remained humble, self-effacing, and obsequious—as if he was apologetic about being so bright and knowledgeable. This, I am sure, Hardy would have found less typical. Unfortunately prophets are even less venerated in their schools than in their countries, and Ramanujan was a perpetual target of the practical jokes of his class fellows. They had little reason to like his preference for solitude and his single-minded absorption in mathematics. Moreover, he was a favourite of the teachers and the headmaster who liked his quietness and, even more so, his ability to turn out school timetables at short notice. Such situations always invite the hostility of one's classmates. In Ramanujan's case, they must have liked even less his interest in the Sanskrit classics, the sayings of saints, and the *puranas*, all of which he frequently recited in the traditional sytle of Brahmin pandits at the least provocation.

[32] C. P. Snow, 'G. H. Hardy', *Variety of Men* (New York: Scribner, 1967), pp. 21–61.

[33] G. H. Hardy, *A Mathematician's Apology* (Cambridge: Cambridge University Press, 1941).

Of these traits, the early loneliness of an only child and search for solitude in latency or pre-adolescence are particularly noteworthy. On the one hand, they are consistent with the developmental histories of a number of creative mathematicians and physicists. On the other, they seem inconsistent with the link between a dominant mother, deep involvement with social realities, and entry into the social rather than natural sciences observed in the West. Perhaps the combination was possible in the greater Sanskritic culture with its studied indifference to boundaries between the arts, the humanities and the sciences, and between the 'tougher' sciences of nature and the 'softer' sciences of society. After all, is not nature (*prakriti*) in this culture the active, mobile, feminine principle of the cosmos, to be grasped instinctively and intuitively on the basis of what one has incorporated of one's earliest and closest experiences of femininity? Is not culture itself associated with the concept of a primordial, 'passive' but authoritative maleness (*purusha*) from which one maintains a manifestly affectless distance as from one's earliest male authority? We shall have to come back to these strange questions again.

Some biographers mention young Ramanujan's fascination with zero and the startling questions he asked his school teachers. (For instance, at the age of twelve, when told that any number divided by the same number equalled one, he asked if zero divided by zero would also be one.) In his classes, he was perpetually involved in mathematical calculations and already in his latency he could give the values of $\sqrt{2}$, pi, and e to any number of decimal places. At twelve or thirteen, he is even said to have discovered the relationship between circular and exponential functions and was very disappointed to find out later that Leonhard Euler had already discovered them.

As mentioned before, this mathematical precocity co-extended with a serious interest in astrology, numerology, occult phenomena, and prescience. The simultaneous interest in modern and occult mathematics might have grown out of Ramanujan's early interpersonal relationships but now it was strengthened by his unconventional exposure to modern mathematics which he had begun to study on his own. This exposure gave him access to the text of contemporary mathematics but not to its culture. This culture might have exerted some internal pressures on him to cut himself off from the more magical systems of ideas; whereas his imperfect socialization to modern mathematics encouraged him to believe that he could enrich and extend the systems of magical mathematics through modern mathematics and make them even more powerful.

Despite his concern with 'para-mathematics' and his childhood dream of finding the 'ultimate mathematics', Ramanujan did have an uncanny sense of the significance of modern mathematics. He got exposed to it at the age of fifteen by a rather ordinary textbook he borrowed—G. S. Carr's *A Synopsis of Elementary Results in Pure and Applied Mathematics*.[34] He was then in the sixth form. The exposure came at the right moment, for a number of anecdotes suggest that numbers in general, and positive integers in particular, were already becoming, as Littlewood was later to put it, his personal friends. And in this friendship an almost feminine intuition and empathy had already started playing a prominent role.[35] There was only one problem. Not knowing Carr to be a mediocre mathematician and presuming his book was a standard mathematical work, Ramanujan once and for all modelled his style of writing on Carr's work. For the rest of his life he continued to produce what could be called the different volumes of Ramanujan's *Synopsis of Elementary Results in Pure Mathematics*.

Why did the mathematician gradually people his world with those anthropomorphic formuale and numbers, as if to contain all human relations within an affectless, formal, controllable system? The answer lies in the cultural meaning of mathematics to him.[36] Ramanujan's first exposure to mathematics (and to that branch of it in which he was later to specialize, the theory of numbers) was through his mother's astrology. This association was only strengthened by his Brahminic heritage in which mathematics and numbers had been, among other things, magical instruments for propitiating and

[34] According to Hardy and Littlewood, G. S. Carr and his book, *A Synopsis of Elementary Results in Pure and Applied Mathematics* (London: Francis Hodgson, 1880), vol. 1, and 1886, vol. 2, are now completely forgotten, except for the inspiration they gave to Ramanujan. This was the only book available to Ramanujan and the mathematical knowledge summarized in it went no further than the 1860s. See Hardy, *Ramanujan*.

[35] The best example of such intuitive powers is the story of the taxi number, too well known to be repeated. See Snow, *Hardy*.

[36] Compare the apparently affectless, anthropomorphic, obsessive-compulsive mathematics of Ramanujan with the highly emotion-laden anthropomorphic scientific world within which Jagadis Chandra Bose worked. On anthropomorphization as a characteristic of creative persons, see Phyllis Greenacre, 'The Childhood of the Artist: Libidinal Phase Development and Giftedness', *The Psychoanalytic Study of the Child* (New York: International Universities, 1957), *12*, pp. 47–52.

controlling fate or *niyati* and the nature of *prakriti*. They helped in the delineation of ritual or magical boundaries of safety (*gandis*), the choice of auspicious moments (*mahurtas, yogas, tithis, kshanas, lagnas*), the correct reading of indicators (*lakshanas*), and the identification of appropriate fetishes, magical guard plates, and protective charms (*kundalas, kavachas*, etc.). Their function was to extend some control over the two cosmic principles of *niyati* and *prakriti*, both feminine, both unpredictable, both aggressively malefic and benevolently creative at the same time. The assumption was that the maleficence could be contained through rituals, magic and mystic intervention (so that the nurtural and creative aspects of the cosmic powers could be fully released).[37] This lent a divine quality to mathematical creativity and made mathematics the 'highest truth'.[38]

Numbers and their interrelations—arithmetic, geometry, algebra—all were, therefore, parts of a cognitive and mythic order. They constituted a language, the grammar of which could not be formalized without rules that were part-sacred. One need hardly add that when not culturally sanctioned and transmitted, such an order would have seemed remarkably close to what clinicians diagnose as a delusional system. But then, by such clinical standards, every rejected theory or cosmology of science turns out to be in, in retrospect, a delusional system. Yet these are the very bricks with which the edifice of science is built.[39] The difference between magic and science lies not so much in their content, as in their internal organizational principles, methodologies, the permeability of their boundaries, and the justificatory principles they use.

Given that traditionally mathematics in India had a magical–divine connection—as chemistry had in the history of the western world—in Ramanujan's case it performed another important function. Though he knew his mathematics to be a gift of the goddess

[37] It is possible that this concept of individual control over an unpredictable feminine principle, and the aim of gaining power over powerful environmental forces to counter feelings of personal insignificance, gave such a strong push to Indian mathematics in ancient times and has made mathematics and mathematically-oriented disciplines the cutting edge of Indian scientific effort in recent times too.

[38] Ramanujan quoted by Hardy, Trinity Papers, Add. MSS a. 94 (7–10).

[39] Probably no great scientist is either wholly rational or wholly magical; see e.g. John Maynard Keynes, 'Newton the Man', *Essays in Biography* (London: Heinemann, 1961), pp. 310–23; and Michael Polanyi, 'Genius in Science', *Encounter*, 1972, *38*, pp. 43–50.

Namagiri who had blessed him by writing on his tongue, he also claimed that sometimes in his dreams the god Narasimha, in a fantastic reversal of roles, revealed his divine tongue in the form of scrolls covered by complicated mathematics superimposed on drops of blood. On these signs of oral grace Ramanujan founded his self-definition as a scholar and—perhaps more crucially at this stage—as a chosen one in whom mathematics was to become a nexus between legitimately violent, divine orality and an apparently sterilized, two-dimensional but nonetheless nonviolent representation of it; between a powerful, cosmic, feminine principle—represented by a magical mother who was so by virtue of being an exponent of magical mathematics—and the grace of a benevolent god representing protective violence.

None of this should be construed to mean that Ramanujan had any sense of a grand mission or a religious calling at that stage. If he had any such ideas, we do not know of them. On the contrary, it appears that he was still a bright village boy trying to do well in areas in which a poor but educated Brahmin's son should try to do well. He was a good student and, on the basis of his high school performance—he got a first class—secured a scholarship in 1903 for studying F. A. at the Government College, Kumbakonam.

What happened during the next year we do not know. But the teenaged Ramanujan now began, even more than formerly, to see his mathematics as a transcendental abstraction and as a means of isolating his affect, and ultimately himself. Perhaps his adolescent asceticism—serving as a defence against the surging instinctual, notably aggressive and sexual, impulses—had something to do with this. He may have sought in mathematics a disembodied counterpoint to knowledge that related to things physical, interpersonal, and empirical. One indication was his singular horror of physiology which, of the subjects he studied, was the only one concerned with the functions of the body and the biological self.[40] The subject also aroused his deepest anxieties about aggression. Ramanujan hated dissection, particularly the anaesthetization and destruction of experimental animals, and there are accounts of how, shedding his customary shyness, he would sneer at a science that forced one to vivisect in order to learn.

Not unnaturally, Ramanujan failed in the first year examination

[40] Suresh Ram, *Srinivasa Ramanujan* (New Delhi: National Book Trust, 1972). Apart from physiology, Ramanujan's subjects were English, Greek and Roman history, mathematics and Sanskrit.

of his college. A popular Indian myth would have it that he failed in mathematics, but this has now been proved wrong—he actually scored 100 per cent in mathematics. (Perhaps the myth-builders wanted to make his victory over modern mathematics total, but it now appears that he failed in English and physiology.) In any case, he lost his scholarship and wandered about for a couple of months in the neighbouring state of Andhra Pradesh with his notebook full of mathematical work. According to one account, Ramanujan ran away from home at the 'instigation' of a friend, but the details have been washed away by time. Another mentions a six-month-long 'mental aberration'. Again there are no details.

The second possibility must be seriously considered. On two other occasions, Ramanujan showed the same peculiar form of amnesia marked by a sudden loss of contact with the immediate environment and a tendency to 'walk away from it all' as if in a trance. It reminds one of the story by H. G. Wells in which the hero was haunted by the vision of a door in a wall through which he was called into an enthralling Eden, away from the realities of the world of competition, ambition, achievement, and success. Wells' hero lost the road to this alternative world in the welter of his myriad transactions with day-to-day existence and finally paid with his life to regain his utopia. Ramanujan perhaps found a partial clue to it in his mathematics, which promised an alternative gateway to this lost world.

To continue: Ramanujan after a while returned home and rejoined the Kumbakonam College. But he could not make up, it is said, for his poor attendance at college and failed to get his F. A. certificate. Still eager to get a degree and prodded by the anxiety of his lower middle-class parents, Ramanujan then joined a well-known college at Madras. The change did not help. He failed again at the Madras University examination in 1907. He now abandoned formal studies altogether, and devoted himself entirely to the private study of mathematics.

For three years he lived in total obscurity and filled up his now-famous ledger-like notebooks, which he had maintained since the age of fourteen, with a mass of mathematical formulae, based mainly on intuition and crude induction and almost invariably without any proof. This stretch of intense activity was interrupted only by his marriage in 1909.

Ramanujan's failure to get a degree and his 'futile' absorption in mathematics had deeply disappointed his parents and, to cure their own 'depression' and the son's 'waywardness', they had recourse to that time-tested Indian psychotherapy: an arranged marriage. The son's lack of employment rarely discouraged fond Indian parents from fulfilling their parental duties in this area and in Ramanujan's case too, his mother soon enough found a bride in Janaki Devi. Since it was an arranged marriage Ramanujan had not even seen Janaki before she landed in his life as his bride. Apparently her cultural background was roughly similar to Ramanujan's.[41]

Janaki was only nine at the time of her marriage. Her parents exercised the option, which Tamil custom gave them, of keeping her with them until puberty. Perhaps Ramanujan's joblessness had something to do with the decision; he was then partly earning his living by coaching students. There was also reportedly some unpleasantness at the time of the marriage; it might have prompted Janaki's father to refuse to formalize the marriage for a while. All we know is that, until 1913, Janaki paid only occasional visits to her husband at Kumbakonam.

Strange though it may seem in view of Ramanujan's ambivalent relationship with his mother, the marriage worked out well. Not only was the husband–wife relationship close by all accounts, which might have been a remnant of the emotional cross-investment between Ramanujan and his mother, but the scientist also developed with his wife—whom he called his 'house'—a low-key, shy but, by the standards of their culture and time, a romantic relationship. That his wife did not stay with him for more than three years in all might have contributed to this. For she was—and is—in many ways a formidable woman. One cannot rule out the possibility that, had they lived together longer she would have triggered some of the same anxieties in her husband that his mother frequently did.

As it happened, Janaki permitted her obsessive husband a lot of room for rumination and isolation and tried to provide him with non-demanding, non-controlling succour. She also perhaps helped him cope more successfully with his ambivalent strivings for autonomy, by taking care of his dependency needs which the partial separation from his mother due to marriage may have heightened. On the other hand, by openly resenting Komalatammal's dominance over her son, Janaki disrupted her mother-in-law's low-

[41] However, in one respect, Janaki belonged to a more conservative family. At the time of marriage she was totally illiterate, whereas even Ramanujan's mother had some education.

key authoritarian control over the household. This might have threatened the defences of her husband, bringing out his latent ambivalence towards his mother. It might also have mobilized the other second-class citizens in the household to identify with Janaki. Srinivasa Iyengar, according to some, became extremely fond of her.

The kind of husband–wife relationship Ramanujan and Janaki developed is thrown into relief by their cultural background. This background demanded undemonstrative conjugality, under-emphasis on sexuality in marriage, and stress on nurture and succour. The underlying institutional need was to discourage within-family fissures developing along the boundaries of nuclear units within joint households. But Ramanujan's relationship with his wife seemed to take on, even while conforming to the institutional demands of an Indian family, the features of a genuine person-to-person relationship. And this within the steel frame of a subculture which had made, in this one area at least, the rules of family-living foolproof over the centuries.

The Ramanujans were a childless couple. Soon after his marriage, Ramanujan was operated upon for hydrocele at Kumbakonam. Given the state of the medical sciences and the medical advice available in those parts, it is not surprising that a short while after the operation Ramanujan started bleeding from his surgical wounds; so the couple were forbidden to have physical relations for about a year. After that Ramanujan went abroad and came back a very ill man to stay with Janaki for only about a year. Thus, it was a case of enforced childlessness. Nonetheless, it is doubtful if the couple could have avoided the repercussions of it. They lived in one of the most conservative sections of Indian society in which the inability to reproduce arouses not only deep feelings of inadequacy and guilt, but also the pity and hostility of others. Pity, because infertility could not but be one of the worst afflictions in an agrarian worldview in which fertility had been historically a powerful cultural motif; hostility, because in a culture with an over-determined concern with the production, preservation, and loss of potency,[42] it sets off one's deepest anxieties about one's own generativity and about the fecundity of nature or of the land that sustains one. The situation is worse for the childless in a linear joint family. The older generation invariably makes clear what it thinks of the couple's, particularly the wife's, inability to

[42] E.g., Carstairs, *The Twice Born*.

supply it with a grandson to perpetuate the lineage and perform the rituals that ensure comforts in after-life. Where life is short and living tough, childlessness deprives the parents and grandparents of even the hope of the pleasures of after-life.

The family, close friends and relatives knew about the operation; others did not. But India's folk wisdom almost always assumes the wife to be the cause of childlessness, unless proof is supplied to the contrary. The family may not have been hostile to Janaki on this score but, to many others, her apparent barrenness may have made her a bad omen and her presence on religious and other festive occasions a source of resentment. It is true that the Iyengar subculture did not openly stigmatize a childless wife as many other Indian subcultures did; nevertheless, in a tradition which valued a woman primarily as a mother, barrenness was seen to be an abdication of womanliness itself.

In such a context, it is not surprising that there were frequent conflicts between the young wife and the centre of power in the family, the mother-in-law. To some extent, the conflicts were inherent in a culture where a major access to social status was through the son. The childlessness of the Ramanujans induced Janaki to make her husband even more the focus of her attention.[43] But the more Janaki tried to find in her husband a compensation for her childlessness and to protect him from the intrusiveness of his mother, the more Komalatammal felt that her son was slipping away from her. As we know, Komalatammal, too, had a history of childlessness. She might have seen in the childlessness of her daughter-in-law reflections of her own past when her own long period of barrenness had made her a target of self-hatred and the hostility of others. Their common history, however, did not lead to any abiding empathy. Instead, both these possessive, powerful and yet insecure women started trying to gain control over the person they had been trying to use as an extension of their selves. Ramanujan contributed to the crisis by trying to remain close simultaneously to his mother and his wife, and in the process arousing the jealousy of both.

Notwithstanding the distractions of marriage and post-marital conflicts, the years 1907–11 were some of Ramanujan's most

[43] See a comparable situation in the life of Jagadis Chandra Bose in Part Two, Section III above.

productive.[44] By the end of the period, he had attained the necessary self-esteem as a mathematician. He told one or two people that he was doing something worthwhile and that Narasimha was personally interested in his work—the first such direct statement of sacred involvement in his mathematics.[45] Some of his acquaintances, too, although not sharing the wide interests of Narasimha, could sense that Ramanujan was on to something important and conveyed their appreciation to him.

But the absence of a proper scholarly setting made his work a risky affair. It was rendered more risky by Ramanujan's belief in private research and meditative mathematics. As a result, most of his discoveries of this period later turned out to be rediscoveries of the work of European mathematicians of the previous century. Being an amateur, he was 'beginning every investigation at the point from which the European mathematicians had started 150 years before him, and not at the point which they had reached in 1913.'[46] Unknowingly, he had pitted himself, as Hardy pointed out, against the accumulated mathematical knowledge of the West and he was bound to lose. However, it was typical of Ramanujan to have independently rediscovered the discoveries of mathematicians of the stature of Euler, Bauer, Georg Riemann, von Staudt, and Adrien Legendre. The Brahmin's mind was a choosy one.

Ramanujan himself found all this out later, when he had built other props to his self-esteem and could shakily cope with his disappointments about his wasted efforts. By that time, in the process of learning about the extent of his wasted efforts, he learnt something about his achievements too. After all, as E. H. Neville and Hardy often said, in areas that interested him, Ramanujan was 'abreast, and often ahead, of contemporary mathematical knowledge'. But, on the whole, by 1911, his early poverty and isolation had already done the damage and he was never able to make up for the loss.[47]

[44] One wishes that more was known about this crucial period 1907–11. Unfortunately, Ramanujan's shy reclusiveness did not allow him to share his experiences of this period with any one.

[45] Radhakrishna Ayyar, quoted in Ranganathan, *Ramanujan*, p. 73. According to this account, Ramanujan believed that Narasimha had directed him in a dream not to publish his results at that stage, because better opportunities would come his way in a short while.

[46] E. H. Neville, quoted in Ram, *Ramanujan*, pp. 25–6.

[47] Hardy always claimed that it was not Ramanujan's early death which was the greatest tragedy of his career—it was rather his inability to get a good mathematical education in the most formative years of his life.

At the end of this period, in 1911, Ramanujan published his first paper.[48] He was then twenty-three. Scholarship of any kind was *ipso facto* valued by the Brahminic literati, and especially by its Tamil variety, and Ramanujan became a slightly better-known figure in the scientific circles of Madras. Within a year he published two more papers in the same journal. He also found a job in the office of Madras Harbour as a petty clerk in 1912. This ensured him a paltry but stable income of twenty-five rupees a month and he continued his mathematics as a hobby for another year. However, he still had to write his notes on packing paper or in red ink across papers already written upon in blue ink.

He also found a kindred spirit in the manager of the Madras Port Trust, Narayana Iyer, who had formerly taught mathematics in a college. He encouraged Ramanujan in his academic ventures, but also injected him with the fear that westerners would plagiarize his work, specially if he gave the proofs of the theorems he was producing. From then on, Ramanujan paled whenever any westerner so much as opened his notebooks and he later carried a part of this suspiciousness to England. As a result, a large part of his earlier work remained a jigsaw puzzle to mathematicians for many years.

By this time he had come in touch with a large number of people and many of the early first-hand impressions of Ramanujan were collected during this period of his life. Most people remembered him as short, plump, light brown in complexion with a high forehead and a big head. On these were fitted a square face, broad nose, and a pair of small ears. A thick growth of hair brushed sideways and the occasional ritual white and red marks put on his forehead completed the picture. Some acquaintances found him uncouth, unshaven, and dirty, and with a stiffer and more feminine version of a Chaplinesque gait. The dirtiness and unkemptness could have been due to poverty, because he made a somewhat different impression on some of his English friends later on.

Two features of his physical appearance seem to have impressed his acquaintances most. The first was his captivating, almost bewitching, eyes. They were the eyes of a mystic and a fanatic, made the more striking by his otherwise shy, withdrawn, rustic appearance and manner. At first meeting he also seemed insensitive and callous to many. But once he turned to his favourite

[48] This and four other papers published by Ramanujan in the *Journal of the Indian Mathematical Society* before he left for England have been reproduced in Hardy, Seshu Aiyar and Wilson, *Collected Papers*.

subject, the mystic significance of mathematics, he emerged from his private world, shedding his customary shyness. His eyes then burnt with an unmistakable fire. Often, in his most intense moments, only the white of his eyes could be seen. His prodigious single-minded industriousness fed into this image of a privileged fanatic. Ramanujan made clear to everyone who came in touch with him that mathematics was the centre of his life. It was not a matter of professional choice or of nationalism or a step to worldly status, but an end in itself and a part of his self.[49]

III

The real break for Ramanujan came at the beginning of 1913 when he was twenty-five. That year, encouraged by two of his Indian patrons, he sent the 'bare statements' of some 120 theorems, mainly formal identities, to the well-known Cambridge mathematician, G. H. Hardy (1877–1947). This was the first step towards the now legendary collaboration between the two mathematicians and perhaps the most remarkable East–West scientific collaboration we have known.[50]

At the time he wrote to Hardy, Ramanujan was still working for the Port Trust in Madras and his salary was still a redoubtable twenty-five rupees per month. But he was no longer an unknown, half-educated villager obsessed with mathematics. His papers had already been published in the *Journal of the Indian Mathematical Society*, and people had begun to call him a genius. However, this estimate of his work in his new social circle had not yet seeped into Ramanujan. Though he had the self-assurance born of a deep faith in the mystic significance of his mathematical work and his 'chosen' status, he felt diffident and insecure confronting the world of academic mathematics.

So he sent his work to Cambridge with a covering note tinged with both obsequiousness and bravado.[51] The mix was dangerous.

[49] The one informant who spoke of nationalism as a major inspiration of Ramanujan actually seemed to have in mind the collectivity of the Indian scientists of the period, rather than Ramanujan alone.

[50] Hardy was the third person to receive samples of Ramanujan's work; two English mathematicians had earlier returned them without comment. Littlewood says with some relish that these two mathematicians, whom he identifies only as Baker and Hopson, felt rather foolish afterwards. For a lively description of the discovery of Ramanujan, see Snow, *Hardy*.

[51] The entire letter has been reproduced in Ram, *Ramanujan*, pp. 22–30. It is

It was bound to make mathematicians suspicious, particularly someone like Hardy who was, by nature as well as by conviction, a sceptic. In 1913, he was thirty-seven, famous and cynical. As a professional, he was only too aware of the mathematical cranks who abound in all parts of the world and pester famous mathematicians. Moreover, Ramanujan's talents were not so obvious as they were to appear to Hardy and others in retrospect. It is now known that two accomplished Oxbridge mathematicians, E. W. Hobson and H. F. Baker, had previously returned Ramanujan's works without comment.

Thus it is not surprising that Hardy was at first suspicious. Rather, it testifies to his personality resources and intellectual acumen that despite his initial doubts he assessed his Indian correspondent's worth correctly within a day. That momentous day has been described by many. Here is the argument which helped Hardy finally to make up his mind:

> I should like you to begin by trying to reconstruct the immediate reactions of an ordinary professional mathematician who receives a letter like this from an unknown Hindu clerk....(some of the theorems Ramanujan sent) defeated me completely; I had never seen anything in the least like them before. A single look at them is enough to show that they could only be written down by a mathematician of the highest class. They must be true because, if they were not true, no one would have had the imagination to invent them. Finally (you must remember that I knew nothing whatever about Ramanujan and had to think of every possibility), the writer must be completely honest, because great mathematicians are commoner than thieves or humbugs of such incredible skill.[52]

Immediately afterwards, Hardy began trying to bring Ramanujan to England with the help of Neville, mathematician and Fellow of Trinity College, who was on a visit to Madras at the time. Others in India, too, were trying to get Ramanujan a fellowship and some money to visit Cambridge. As a result of these efforts, he received

---

noteworthy that Ramanujan in his first letter to Hardy claimed to be 23 years old, when he was actually 25; only a few weeks afterwards Gilbert Walker, a mathematician visiting Madras, was told that Ramanujan was 22 years old; ibid., p. 27. It is not known whether Ramanujan's friends encouraged him to misrepresent his age or whether he himself thought of this ploy to impress his prospective patrons.

[52] Hardy, *Ramanujan*, pp. 1–3.

within a year both a fellowship tenable in England from Madras University and an invitation from Cambridge. However, his first response to the invitation was a disappointment to his patrons. He refused to leave India because of his caste prejudices. Crossing the seas was considered polluting by many conservative Hindus and Ramanujan on such issues was a conservative.[53]

His mother's adamant opposition to the idea, based on fears of pollution, was another source of external—and internal—resistance.[54] Fortunately, she withdrew her objections later, after most appropriately dreaming that Namagiri, the family goddess, had interceded in favour of her son's journey to the West. The event was less esoteric than it may seem at first. Convenient dreams had played a major role in Ramanujan's life. As in the case of many of India's major folk heroes, his dreams, too, often sanctioned his participation in things that were novel or implied the defiance of conventional authority.[55] There was also another development within the family which might have had something to do with the dream. Ramanujan's teenaged wife, Janaki, had come to stay with her in-laws permanently a few months earlier. Perhaps the prospect of her son being away from her daughter-in-law cum rival Janaki helped bind Komalatammal's anxiety about being separated from her son.

Ramanujan reached Cambridge in April 1914 and stayed the first two months with Neville. He must have been the oddest of arrivals at the university, with his ill-fitting clothes, newly cropped hair—he had cried like a child in Madras when his sacred tuft was removed before his departure for England—and fiercely tight new shoes. And his discomfort showed. Long after, some of his English acquaintances were bitter about his being made to suffer all this merely to look more like an English gentleman. But in 1914 Victorian England was still a living reality. And the insistence on formal dress was certainly not diminished by the heightened nationalism that was to plunge Europe, within a few months, into another massive orgy of violence.

A place was found for Ramanujan at Trinity College and the

[53] Perhaps there was also the painful memory of his first 'departure from home' for Madras.

[54] Characteristically, Ramanujan's father did not interfere in the matter at all, nor was he a party to the decision.

[55] Ibid. It is possible that these convenient dreams in Indian life and mythology are homologues of what Freud called 'obliging dreams'. See Sigmund Freud, 'Psychogenesis of a Case of Homosexuality in a Woman', in James Strachey (ed.), *The Standard Edition of the Complete Works of Sigmund Freud* (London: Hogarth, 1955), *18*, pp. 165–6.

task of educating him fell to Hardy and Littlewood. They were conscious that this re-education had to proceed without destroying the self-esteem of their ward which the latter's extreme modesty suggested was fairly vulnerable. But the Indian was less fragile than he looked: the inner strength that had seen him through his Madras days was very much a part of his personality. His English patrons, one suspects, were taken in by his diffident manner. Littlewood today admits this indirectly; Ramanujan, he says, was too engrossed in his own work to learn much. Even Hardy, at one level, knew this. In retrospect, he felt he had learnt more from Ramanujan than Ramanujan from him.[56]

It did not take the strange Indian long to become a great academic sensation and a legend in Cambridge. In 1916 he received an honorary B. A. degree from the university, and in 1918, at the age of thirty, he was made a Fellow of the Royal Society and of Trinity College. It is doubtful if these formal rewards helped him gauge his new position in the world of science. Socially, he remained as reclusive and oblivious of his environment as ever. And he continued to shock and embarrass his friends by deciding to apply for undergraduate prizes at a time when he was already being compared to Euler and Karl Jacobi. The attitudes evident in the pathetic applications he wrote in India during 1910–13, seeking a job and soliciting some attention for his mathematical work, had persisted in spite of acclaim and success. One would imagine that though Ramanujan knew the value of his work, he did not know its price. At least, he scarcely understood the academic status system of which he had become a formidable if innocent part.

The most remarkable aspect of Ramanujan's encounter with the West was however his relationship with Hardy. Neither of them were the same after meeing the other, particularly Ramanujan, who had for the first time the chance of living the life of a creative mathematician in his own right, rather than remaining a cute example of a prodigiously endowed natural genius in mathematics. Hardy gave him exactly what he needed: non-intrusive nurture and 'unemotional' support. No other break Ramanujan had had could compare with this. Littlewood is right in saying that his 'genius had this one opportunity worthy of it'.[57]

[56] Quoted by Wilson, Trinity Papers, unnumbered.

[57] Littlewood, *A Mathematician's Miscellany*, esp. p. 90.

Were the bonds between Ramanujan and Hardy solely intellectual and scientific ones? How did the differences in their cultural backgrounds and personalities influence their relationship? Did the fact that one of them belonged to a subject society and the other to a powerful imperial power at the peak of its glory matter to either of them?

Their mathematical gifts were certainly to some extent complementary. I can only refer the reader to Littlewood's fascinating account of the way a joint paper was written by Hardy and Ramanujan.[58] It is more difficult to speculate, from this side of history, on the type of interpersonal dyad built by the two collaborators coming from two antipodal cultures. But a few guesses can yet be made.

The son of a modest middle-class family of teachers, Hardy always felt an outsider within the British élite culture of which Oxbridge was the ultimate academic symbol. Contemporaries recall his partiality for non-white students and colleagues, and his shyness and diffidence in dealing with the progeny of the English upper class these universities were crowded with.[59] He probably saw in Ramanujan aspects of his own self: a marginal man fighting tremendous odds and promising to upset the steady applecart of British academia. The Indian provincial, a colonial subject and, like him, a shy, introverted outsider and an underdog, was the mightiest weapon Hardy could have discovered to attack the Establishment with.

The awe-inspiring commitment of Ramanujan to mathematics must have inspired a special respect in Hardy. Hardy, his autobiography explicitly states, never had any intrinsic commitment to mathematics to start with.[60] Along with the advantage of 'mathematically minded' parents, what he grew up with was a substantial quantum of mathematical talent. This was shaped to perfection by his socialization as a scientist and his own sharp perception of the differential advantages of a mathematical career. It is this calculated occupational choice that helped Hardy develop his abiding

[58] Ibid.

[59] This only apparently contradicts Snow's description of Hardy as a scintillating conversationalist. His conversations were hardly expressions of interpersonal warmth or attempts to relate to people. Characterized by a 'grammar' derived from cricket, and tinged by a tendency to shock, they were virtuoso performances which appear to have been desperate attempts to formalize and bind anxiety, even depression.

[60] Hardy, *A Mathematician's Apology*.

allegiance to the discipline. However, this was the allegiance of a professional. On the other hand, Ramanujan had the commitment of a man possessed. Even within his own world, he saw himself as a mystic and a *yogi*, not as a Brahminic preceptor or *acharya*. Such self-transcendence through knowledge a self-conscious rationalist like Hardy could only admire but never duplicate.

Hardy and Ramanujan shared another marginality which I can indicate only imperfectly. One of the distinctions between pure mathematics and the other sciences is the 'restrictions' that pure mathematical creativity imposes on the culture of the discipline. A mathematician can himself subscribe to the dominant style of science in the modern West with its positivist emphases on manipulation, control, prediction and power. But such a style is essentially incompatible with the 'natural' of pure mathematics, which perforce stresses intuitive gestalts and certain aesthetics of form—what Poincairé calls 'the beauty of equations' in another context. Hardy, on the other hand, 'had no faith in intuitions, his or anyone else's.' Yet his writings give one the feeling that at some plane he was sensitive to the importance of the other culture of mathematics with which Ramanujan was linked and which had become recessive at the time.[61]

I doubt if this sensitivity made Hardy conscious of the analogous stylistic differences between the western and the eastern sciences. For Hardy, there was no 'Indian science'. For all his marginality, for him the only science was the one to which he was socialized. Nor could he distinguish between the culture of science, as something that could be parochial and ideologically coloured, and the formal text of science, as something more universal and objective. Nonetheless, Hardy perhaps saw in Ramanujan a personification of the speculative, intuitive, and aesthetic elements which, although recessive in the ultra-positivist culture of western science, were the very stuff of pure mathematics. Hardy's whole upbringing, his professional identity and the scientific norms he had internalized, must all have protested against any admission that

[61] To give an example, he, like Littlewood, sensed that Ramanujan was a mathematical anachronism for two reasons. First, Ramanujan's *forte* was equations and the day of equations, Hardy felt, was over. Second, Ramanujan's weakness was the technology of proof, which had become central to the discipline in his time, and which, as Hardy knew and said, was a relatively inferior and mechanical part of pure mathematics. One suspects that Hardy had at least an intuitive awareness of Ramanujan's self-confident loyalty to the other tradition of mathematics.

such a philosophy of science was possible or legitimate. But he could pay symbolic homage to this philosophy by cultivating and nurturing Ramanujan.

Apart from the fact that his style was closer to the culture of natural philosophy than to that of the experimental sciences, Ramanujan also represented a third culture of science. Early in life he had learnt to use mathematics as an instrument of magical power, extra-sensory perception, and astrological prediction. This use of mathematics could not but arouse anxiety in many scientists because, as they liked to believe, it was on the ruins of this third culture that the modern sciences had been built. Hardy at least consistently denied that Ramanujan had any tendency to equate mathematics with magical intervention in nature and society. One guesses that Ramanujan's 'superstitions' unnerved the aggressively positivist Hardy because they made Ramanujan look irrational—Hardy's identification with his friend was too deep for him to allow that the Indian was anything less than totally rational—and seemed to affirm that the world of mathematics was not as sterilized as the mathematicians would like it to be. In addition to those non-demanding, controllable, and compulsively rule-bound numbers and symbols, Ramanujan's eccentricities seemed to say, that world, too, had its passionate demons.

Thus, through an involved process of intellectualization, Hardy developed a blind spot which, instead of weakening, cemented his bonds with Ramanujan. Hardy was an atheist in a society which was not only puritanic but also conformist. He gradually came to believe that his protégé was an agnostic too. This belief was based on a single statement of Ramanujan's, which Hardy recounted. 'I remember well his telling me (much to my surprise) that all religions seemed to him more or less equally true.'[62] From this one statement, which is actually part of the prescribed daily prayer of a Brahmin, Hardy concluded that Ramanujan had no definite religious beliefs and that he 'saw no particular good, and no particular harm, in Hinduism or any other religion.'[63]

Anyone with even a superficial acquaintance with Hinduism would immediately see the absurdity of this conclusion. Today, the available social anthropology of the Indian civilization has made it unnecessary to stress that the greater Sanskritic culture, while institutionally one of the most rigid, has always been ideologically

[62] Hardy, *Ramanujan*, p. 4.

[63] Ibid. To get a feel of this Hardy, see his *A Mathematician's Apology*.

one of the most tolerant; that it has always rejected the idea of a chosen people with exclusive claims to revealed truth and always disavowed any monopoly of the technology of personal salvation. These were at the time much less obvious. The possibility that a religion could regard other religions, including various forms of atheism, to be different ways towards the same goal, and could even forswear its claims to a superior revealed truth would have been totally incredible to a down-to-earth western sceptic and anti-cleric like Hardy; he would have rejected the idea outright. It is therefore a reasonable guess that Hardy projected his Judaeo-Christian concepts of religion and religiosity into Hinduism and arrived at an image of his protégé that was more congruent with his own needs. On his part, Ramanujan apparently practised with his naïve collaborator what the latter thought Ramanujan to be practising with his Indian friends: 'a quite harmless, and probably necessary, economy of truth'.[64]

Hardy was an outsider in yet another sense; he had strong homosexual needs. Since this is Ramanujan's story, not Hardy's, I shall provide only a brief outline of the personality background Hardy brought to his relationship with his Indian friend.

Margins are meaningful only with reference to the centres to which they are marginal and every marginality respects its corresponding centrality by carrying within it intimations of the latter. Hardy was a marginal man with a strong touch of strangeness about him. But the strangeness, paradoxically, was of a predictable variety; one might even hazard the guess that it was in some ways promoted by aspects of his society and by his upbringing. His early socialization in a middle-class family—his father was a bursar, his mother a teacher—and in a rough, all-male, typical public school, which he deeply hated and where he once almost died, were later on continuously endorsed by the ready-made, over-defined masculinity and the highly compartmentalized sexual identities of the élite culture of Edwardian England. Perhaps this by itself would have been adequate to gradually ease him into the homoerotic culture that thrived in the British public schools, Oxbridge and Bloomsbury, and received part of its sanction from the neo-Hellenism then in vogue. This culture, though not blatant homosexuality of the type that led to the trial and sentence of Oscar

[64] Hardy, *Ramanujan*, p. 4.

Wilde at the turn of the century, had become, in a sexually repressive society, a major organizing principle of intellectual life. Those living the life of the mind, in turn, accepted the homosexual personality type if not as a preferred model of total dissent, at least as a mode of predictable deviation.[65]

To start with, Hardy's homosexual needs might have found expression mainly in the aesthetism of a remarkably handsome man trying to live up to his self-image of an Adonis blessed with perennial youth. Such a self-image might also have received some indirect validation from the culture of Hardy's discipline, the very essence of which, according to one historian of mathematics, is its 'eternal youth'.[66] But as C. P. Snow's sensitive account makes obvious, Hardy was not merely an individualist trying to get through his overt eccentricity and conventional brilliance what he could not get through his mathematics. He was also a lonely man, shy, deeply self-conscious and fearful of company. He never married and never made any significant emotional investment in any woman except for his sister in his middle age. As far as deep relationships were concerned, he lived in a virtually all-male world. His apparently platonic friendship with Littlewood possibly only hid its libidinal content by being seemingly aphysical.

At the same time, Hardy was a person desperately fighting his loneliness and self-consciousness through smart conversation and self-assertiveness, and a narcissist not at peace with his narcissism. He did not like to be photographed, did not tolerate mirrors at home, and whenever he went to a hotel his first move was to cover up the mirrors. He also was a depressive with strong self-destructive tendencies who never forgot that a mathematician was relatively old by the time he was thirty. The culture of homosexuality at Cambridge and the Bloomsbury 'traditions of higher sodomy', as some have called it,[67] gave him only partial protection against the potency-driven tough-mindedness of the outer society and his own complex, tortured self.

A part of Ramanujan's attractiveness to Hardy lay exactly here.

[65] See for instance Michael Holroyd, *Lytton Strachey* (London: Heinemann, 1967), vol. 1. Holroyd's book provides excellent material on the dominant interpersonal style at Oxbridge at that time.

[66] T. Bell, quoted in Ram, *Ramanujan*, pp. 82–3. For identifying the pattern of Hardy's interpersonal relationships, I have partly depended upon our interviews with people who knew him. Some hints are also available in Leonard Woolf, *Sowing* (London: Hogarth, 1961), pp. 110–13.

[67] P. Levy quoted in D. E. Moggridge, *John Maynard Keynes* (Harmondsworth: Penguin, 1976), pp. 10–11.

It was probably not merely the Indian's self-transcendence that impressed Hardy, but also his remarkably integrated self. It was as if Ramanujan, taking advantage of his own history and culture, had found an identity in which femininity, as defined by the western culture, was a valued part. Here was a shy, withdrawn man of 'feminine' build and ways, who knew he conspicuously resembled his mother and grandmother in looks and who shared the family myth that he was a reincarnation of his grandmother. Yet, he retained in him a self-acceptance and serenity which Hardy lacked. True, Hardy perhaps found in his collaborator's personality a classic instance of the phallic woman, the ultimate love object underlying the homosexual's search for feminine men. Being a narcissistic homosexual, he probably also found in the Indian a love object similar to himself, by loving whom he could symbolically get love himself.[68] But the main strength of the relationship was, I suspect, Hardy's discovery in Ramanujan, for the first time, a comfortable, non-threatening figure who did not share the insecure, overly masculinized self of the English social élites of the time. Ramanujan, on his part, with his ambivalent feelings towards an interfering and possessive maternal authority and his emotional distance from his father, found in Hardy for the first time an intervening, caring, close male authority who did not trigger his ambivalence towards all combinations of intervention, care, and proximity.

One area of life where Hardy openly showed his femininity was his intense, uncompromising pacifism.[69] Pacifism was not a proof of his homosexuality but an expression of some of those gender-specific identifications his culture and times had forced him to drive underground. It is nearly impossible to convey today the extent to which pacifism and femininity were psychologically intertwined in Edwardian England. Even a casual knowledge of the lifestyle of the English gentry of the time would suggest that the extroversive jingoism and chauvinism that World War I spawned could be negated, at least at one plane, by aggressively affirming one's 'effete' and 'effeminate' intellectual self and by being a militant pacifist. As if, by being a 'queer' professor in a protected

[68] Otto Fenichel, *The Psychoanalytic Theory of Neurosis* (New York: Norton, 1945), pp. 331, 332.

[69] Snow, *Hardy*. Hardy did not join the British war effort and stayed back at Cambridge because of his pacifist views. This made him unpopular in certain circles, but also enabled him to spend the entire 1914–18 period with Ramanujan when almost all important Cambridge mathematicians were away from the university.

environment like that of Cambridge, one earned the right to defy both the English élite identity, crudely summarised in the image of John Bull, and the plebeian's alternative identity of the Tommy.

Thus, Ramanujan's apparently 'ultra-radical' pacifism, as Hardy called it, gave Hardy's dissent some much-needed psychological support. The Indian's pacifism may or may not have had its ideological roots in the deep fears of aggression and the defensive demands for its total unconditional control which characterize child-rearing in even some of the 'martial' communities in India. But it had its practical ramification in Ramanujan's steadfast refusal to work on any problem connected with the war, even when requested to do so by Littlewood.

This linking of homosexuality, aesthetism, femininity, and pacifism also influenced Hardy's attitude to mathematics. He found applied mathematics a 'useful science', and hence, 'repulsively ugly and intolerably dull'.[70] Many defensive mathematicians considered this attitude to be a form of sickening 'cloistral clowning'—an admission of shock which must have pleased Hardy no end. Others tried to blunt the sharper edge of his dissenting views—and the anxiety they provoked—by accepting Hardy as a 'strange' eccentric whose opinions and prejudices were deliberately assumed poses. They read him as a 'predictably odd' professor on show in an academic reservation. (None of these critics grasped that Hardy had somehow sensed the increasing instrumental use of science to gain power and control over man and nature and to express one's destructive impulses.) As it happened, the ideas of science as play and science as self-exploration were not unknown to Ramanujan. His Brahminic world-view refused to overvalue practical knowledge and knowledge not accompanied by introspection.

Behind all this was Hardy's regard for a certain cultural strength that Ramanujan carried with ease and elegance. Ramanujan belonged to a society that demarcated sexual roles much less rigidly than did the modern West. In many situations his society shuffled or switched the culturally-defined gender-role-specific qualities across gender boundaries. It accepted and even valued Ramanujan's pronounced femininity. Such femininity was after all one of the traditional manifests of spirituality, particularly yogic powers and godliness. The man of religion in India was expected to

[70] Hardy, cited in E. C. Titchmarsh, 'Obituary of G. H. Hardy', *The Journal of the Royal Mathematical Society*, April 1950, pp. 81–8.

be rather more bisexual and rather less concerned with maintaining the this-wordly boundary between the sexes. Ramanujan's self-esteem was born of an awareness of this acceptance of his feminine self. To this extent he served as Hardy's target of conflict-free identification, perhaps even as an ego-ideal. Hardy once wrote, 'I owe more to him [Ramanujan] than to anyone else in the world with one exception, and my association with him is the one romantic incident in my life.'[71] He was perhaps not speaking here of a person with whom he had acted out his passions, but of one who had brought adventure and high drama as well as care, responsibility and tenderness into his life.[72] He was indeed paying homage to a person who had seemingly redeemed his troubled sexual identity.

In this congenial, cosy environment Ramanujan's creativity blossomed. His finest papers were written jointly with Hardy at Cambridge, and he seemed well set to lead the serene life of a Cambridge don. Cambridge, particularly Trinity, was tolerant of all kinds of eccentricities and that helped. Ramanujan managed to live there the life of a devout Brahmin without much difficulty, though not without grumbling. He continued to hate the ways of the heathens and, more understandably, the English weather and English food. Even the bedrooms posed problems. Ramanujan once complained to Prashanta Mahalanobis that the extreme cold forced him to sleep in his overcoat, with a shawl wrapped around him. Mahalanobis went to his bedroom to see whether he had enough blankets, and found that the bed had a number of them but all tucked in tightly, with a bedcover spread over them. Ramanujan did not know that he should turn back the blankets and get into the bed. The bedcover was loose, and he was sleeping under it wearing his overcoat and shawl.

In his spare time, Ramanujan read the *puranas* and all sorts of mystical, theosophical, and astrological stuff, attended popular lectures on the Ramayana and the Mahabharata and, despite the English winter, often wore his cotton *dhoti* and shirt, avoiding shoes and socks. He also cooked his own food, convinced that even

[71] Hardy, *Ramanujan*, p. 2.

[72] The other role perhaps only Littlewood could have played in his life. Whether he played it or not will remain a matter of conjecture. Another of those lonely, brilliant bachelors at Cambridge, Littlewood was a more hardy, masculine specimen who reportedly had even had an illegitimate child.

vegetarian food, when ordered from the college kitchen, was polluted. The one place he enjoyed visiting was the London zoo. He talked about it at great length on his return to India.

He also wrote regularly to his parents, assuring them that he continued to observe orthodox practices. But an even better index of his conformism was his explanation of a bombing raid on Liverpool in which he was caught during the war; he considered it to be God's punishment for drinking a glass of Ovaltine, a beverage he subsequently found contains a small measure of powdered egg. He immediately packed up and left for Cambridge to avoid further temptations.

Ramanujan in Cambridge also apparently became more sensitive to any kind of personal rejection—more touchy, as some of his friends were to describe it later. This touchiness was not new in him and some of his older friends at Madras knew about it, but in England it became more pronounced. He was always socially withdrawn; now he became something of a recluse in his rooms in college. However, one need not read too much into this behaviour; first, he had always tried to isolate himself, to immerse himself in his own orderly world of numbers, and to sterilize his inner world peopled by those for whom he carried conflicting feelings. Secondly, what to outward appearances was a search for solitude in a strange place might also have been the fear of a sociality that presumed new norms and ways of social life. There is some evidence in his letters that Ramanujan tended to make emotional investments in a few relationships, mostly with Indians, and to abstract and reify the rest. In fact, goaded by his loneliness, he later began to have excessive expectations of these Indian friends. For instance, some Indian contemporaries recount how at a dinner given by Ramanujan they suddenly found their host missing. He had actually walked straight out of the house and away from Cambridge as he felt that his guests had not been sufficiently appreciative of his cooking.[73] It is not easy to reconcile such sensitivity with the sturdy ego strength he showed as a lonely discoverer in Madras. He survived on a controlled, ego-syntonic split, which allowed him self-confidence

[73] This is the fourth instance of that peculiar tendency to 'walk out' suddenly on a confusing, and perhaps painful, interpersonal world. Apart from the Andhra Pradesh trip and the Liverpool incident, once, in Madras, he had happened to see a man buying a ticket and boarding a train, and as if in a trance, followed suit. Only after a while did he come back to his senses, wondered what he was doing alone in a train, and returned home.

in one sector and self-doubt in others.[74] Perhaps mathematics was his one conflict-free sphere which remained relatively unaffected by the anxieties he lived with.

The perimeter of Ramanujan's interpersonal world shrank not only due to his fear of being hurt. He had no time for any of the facilities available in the college which could have brought him in touch with other scholars. He did not teach, was not interested in sports (Hardy, for instance, was not able to infect Ramanujan with his passion for cricket), and there is no evidence of his ever having gone to plays and shows (except once when he saw and immensely enjoyed the comedy *Charlie's Aunt*) or interesting himself in the activities of the Indian Majlis (that delightful little club where succeeding generations of scions of the Indian aristocracy picked up their nationalism and radicalism, and imported these, along with their English accents, degrees and blazers, to the seller's market called India). What surely cut off Ramanujan most effectively from other scholars were his food habits. He never dined in college, where he would have encountered other Fellows.[75]

Only in one respect did Ramanujan try to move out of his loneliness and flout his 'culture'. He tried exchanging letters with his wife. Unfortunately, this adventurous assertion of autonomy proved costly. One of those routine quarrels was on between Janaki and her mother-in-law, and Ramanujan's letters, as well as Janaki's to him, were intercepted and destroyed by Komalatammal. South India fifty years ago was not renowned for respecting the privacy of conjugal communication. Nevertheless, even by the standards of that society, a wife could rightfully share with her in-laws letters received from her husband, and a husband too had the right of access to his wife's letters. Yet nobody in the family, certainly not Ramanujan, protested against the manner in which Komalatammal's authority was exercised in this instance. One can only construe it as yet more evidence of the mother's immense power in the family and the son's relapse into conformism after the defiant gesture of writing to his wife.

[74] On segmentalization as a characteristic adaptive mode in the Indian personality, see e.g., Milton Singer, *When a Great Tradition Modernizes* (New York: Praeger, 1972).

[75] This however had its brighter side too. Littlewood has said in another context, 'the thing to avoid, for doing creative work, is above all Cambridge life, with the constant bright conversation of the clever, the wrong sort of mental stimulus'. Littlewood, *A Mathematician's Miscellany*, pp. 69–75.

But Ramanujan suffered all the same. Probably his strong defences against his latent anger towards his mother were now to some extent breached, arousing guilt and self-hatred. What his friends noticed were touches of depression and even melancholia. For many of them this was evidence of homesickness;[76] they had no reason to guess that, in a person so clearly a mother's son, such depression was likely, at some point, to find expression in self-destructive behaviour.[77] They did not know that the thought of death had exercised an eerie fascination over Ramanujan from an early age and that he had shared the astrological 'knowledge' with his mother that he would die young—a prophecy which was, as it turned out, a self-fulfilling one. Nor were these friends sensitive to the fact that in his culture one of the time-worn techniques of expressing anger against one's 'significant others'—and a defence against the moral anxiety generated by such anger—had been to turn upon one's own self.

Other things too may have happened during the year. Circumstantial evidence suggests that it was during the years 1916–17 that Ramanujan came to appreciate for the first time the full magnitude of his wasted efforts.[78] It had gradually become clear to him, mainly due to his exposure in Cambridge, that it was not a matter of a theorem here or an equation there, but that about three-fourths of his earlier work done in India had been a mere rediscovery. He also probably guessed that, given the relationship between physical age and mathematical gifts, he stood no chance whatsoever of making up the lost time. His early poverty had already done its bit. It is true that in most transcendental theories of knowledge, glory attaches as much to those who go through an experience and enrich themselves as to those who cull from the experience externally valued, objective discoveries. It is also true that Ramanujan's culture frequently valued a thing because it was produced by someone in particular, and not a person because he had produced something

[76] Hardy knew better; there is indirect evidence that he tried to mitigate Ramanujan's anxieties by bringing about a rapprochment between Ramanujan and his family. See his letter to S. M. Subramanian, 20 September 1977, in Srinivasan, *Ramanujan*, pp. 69–75.

[77] Moreover, one unconscious motivation for suicide is said to be the suicide's mystical oceanic longing for union with the mother. See Fenichel, *The Psychoanalytic Theory*, pp. 400–1.

[78] The first inkling had come during 1914–15. See, for instance, his letter to Krishna Rao, S. M. Subramanian and S. Narayana Iyer, in Srinivasan, *Ramanujan*, pp. 13–27, 29, 32–3.

in particular. Nonetheless, it must have been painful for him to know that partly he was being adored as a grand eccentric rather than as a person whose work was a benchmark in the history of world mathematics.

All this resulted in a deep personal crisis, and some time in the second half of 1917 Ramanujan attempted suicide by jumping on the path of a train on the London Underground.[79] He escaped death narrowly, but was badly injured. An attempt to commit suicide was a penal offence under British law and Ramanujan was duly picked up for questioning. It was the shy, retiring Hardy who saved him from gaol by bluffing the investigating police officer.[80] Hardy himself had strong suicidal tendencies, expressed later in an unsuccessful and pathetic attempt at suicide and one can well understand the empathy between him and the lonely Indian fighting depression. Ramanujan too confided in Hardy some of the personal problems which had prompted the attempt. Neither his friends nor his family came to know anything of the incident.[81] Hardy was now persuaded that his friend's family had a role, however trivial, to play in the history of mathematics. He began taking a more active interest in the family quarrels of the Iyengars, and once even agreed to act as arbiter.

The attempted suicide should not blind us to the working relationship Ramanujan developed between his worldview and his science. His problem was not with the text of his mathematics, with its philosophical implications, or with the relationship between his field of knowledge and the world in which he lived. And he knew this. After all his greatest professional tragedy, namely his lack of a sound mathematical education, was 'external' to him, according to both Hardy and Littlewood. He was deeply unhappy that much of his work consisted of what could only be called rediscoveries, but it was not the unhappiness of a person who had lost in professional competition. He had a philosophy of life, and the mathematics he

[79] The attempted suicide was first mentioned in public in India fifty years after the event. And the mention immediately provoked protests. See a brief discussion of Indian attitudes to history in the 'Introduction'.

[80] Hardy told him that Ramanujan was an FRS; the officer seemed duly impressed. The fellowship actually came to Ramanujan after some weeks. Years later Hardy learnt that the officer knew this all along, but nonetheless had wanted to be helpful.

[81] Years after the death of Ramanujan, S. Chandrasekhar, the astrophysicist, told Janaki Devi for the first time the cause of the marks on Ramanujan's knee about which she had been worried throughout his life.

knew and the mathematics to which he had exposed himself formed an integrated whole from the point of view of that philosophy. He even knew his priorities. His friend P. C. Mahalanobis once said, 'He [Ramanujan] would have been better pleased to have succeeded in establishing his philosophical theories than in supplying rigorous proofs of his mathematical conjectures.'[82]

What was the content of the self-sufficient relationship Ramanujan forged between his work and environment? Mahalanobis' reminiscences are again pertinent, not only because he knew Ramanujan at Cambridge, but also because as an agnostic, Marxist, mathematical statistician, he can be expected to ignore the magical interpretations which adulatory biographers like Ranganathan try so hard to foist on us. This is what Mahalanobis, the Indian modernist, says about the mathematician, against Hardy's testimony:

> He was eager to work out a theory of reality which would be based on the fundamental concepts of 'zero', 'infinity' and the set of finite numbers.... He sometimes spoke of 'zero' as the symbol of the absolute (*Nirguna Brahman*) of the extreme monistic school of Hindu philosophy, that is, the reality to which no qualities can be attributed, which cannot be defined or described by words and is completely beyond the reach of the human mind; according to Ramanujan, the appropriate symbol was the number 'zero', which is the absolute negation of all attributes. He looked on the number 'infinity' as the totality of all possibilities which was capable of becoming manifest in reality and which was inexhaustible. According to Ramanujan, the product of infinity and zero would supply the whole set of finite numbers. Each act of creation...could be symbolised as a particular product of infinity and zero, and from each such product would emerge a particular individual of which [the] appropriate symbol was a particular finite number.[83]

The social anthropologist may call this only another instance of the ritual neutralization of western science, but it does not need much imagination to guess that, in Ramanujan's life, this also was a search for a state which reduced persons and social relationships to abstractions and contained them within an obsessively ordered universe of ideas. It was an attempt to use mathematical abilities to symbolically freeze the living universe of people into a semi-magically controlled world of numbers.

[82] P. C. Mahalanobis quoted in Ranganathan, *Ramanujan*, p. 80.

[83] Ibid., pp. 82–3.

In May 1917, while still in England, Ramanujan was found to have tuberculosis. Probably the early symptoms had been neglected, for it became obvious after the diagnosis that the illness was at an advanced stage and, given his Brahminic food habits and the existing state of medical knowledge, incurable. At best, drugs, nursing, a careful diet, and change of climate could buy him some time. But even this seemed unlikely because he was unwilling to live in a suitable environment under proper treatment.[84] He placed more reliance on a *kavacha*, a magical guard plate, to protect himself.

His diet was another problem. He had always been finicky about food and dismissive about western food. The illness made things worse. His correspondence with some Indian friends, particularly letters he never expected to be shown to others, reveals that it was not merely a matter of abjuring animal proteins or an inability to adjust to English cooking. It was not even austerity; Ramanujan apparently enjoyed good Indian food, whereas he could not even eat vegetarian western food, such as cheese, bread, butter or jam (in spite of his misleading pet name in Cambridge, 'Dear Jam'). One suspects that, as with the western lifestyle as a whole, English cuisine too was entirely outside his frame of reference.[85] He was a culturally self-sufficient man with strong psychological defences and neither his environment nor any form of rationality except his own had easy entry into his life. On the other hand, though he did not seem to recognize it, his culinary skills were less formidable than his mathematical ones. The price he paid for his failure to realize this, even after he had tuberculosis, proved heavy.

[84] On how difficult a patient Ramanujan was, see Francis Dewsbury's letter to G. H. Hardy, 22 December 1919, Trinity Papers. Hardy suffered too. Once in a London hospital, when cucumbers were not in season, Ramanujan wanted cucumbers, of all things, to eat. Hardy got him some, nobody knows from where.

[85] See Ramanujan's letter to his friend A. S. Ramalingam, 19 June 1918 and Ramalingam's reply of 23 June 1918, MSS a 94 (1–6), Trinity Papers. Also see Ramalingam's letter to Hardy of 23 June 1918, ibid., which suggests that Ramanujan would eat plain boiled rice sprinkled with red pepper or hot pickles rather than any English vegetarian food.

One should, however, remember that if Ramanujan's fads tell us something about the distinctive pattern of his orality because they centred round food, they also underline Ramanujan's obsessional and compulsive defences. True, they were included by his culture within the range of normal adaptive responses; but the fact that certain behaviour patterns are considered normal in some cultures and abnormal in others does not, by itself, deprive these patterns of their psychodynamic content.

It was pointless now to linger on in England. Ramanujan himself, his friends and doctors, all agreed that he should return to India. At some level they all knew that the death sentence had been passed and it was better to be close to one's relatives and to one's 'home'.

Did he during his last days in England regret having to leave? Or ponder over the question which hounds his biographers even now, whether his trip to England was justified? Hardy was sure that the trip was necessary and Ramanujan would have died an unhappier death if he had not moved out of India.[86] Perhaps he was right. But did Ramanujan look at it that way? It is true that he had vaguely thought of returning to England and left behind some of his personal belongings. But must he not have felt in his lonelier moments that the uterine warmth within which he had pursued his mathematics during the early part of his life was, if not more rewarding, at least more real than the alien world of Cambridge? The answer may well be 'yes'.

The semi-rural Kumbakonam, after producing the only great man in its long history, has relapsed into somnolence, assimilating Ramanujan into a local legend only partly tainted by life. Similarly, to the shy, provincial Brahmin who had made good in the company of the great and was now on his death bed, his five years in England must have seemed as ephemeral and mythical as any memory of grandeur. His English sojourn had merely torn him asunder from his gods, his way of life, and his simplicity and autonomy. In the process, science may have gained, but he had certainly lost.

IV

Ramanujan returned to Madras in March 1919. At first he was happy to be back with his wife, his relatives and friends, and his beloved Tamil food. Though bed-ridden, he regularly received

[86] I doubt if one can say as confidently as Hardy did that Ramanujan's English trip made him, professionally, what he was. One wonders if it ever struck Ramanujan that out of the roughly eight areas in which he worked (hypergeometric series, partitions, definite integrals, elliptical integrals, highly composite numbers, fractional differentiation, and number theory) it was his work on fractional differentiation which perhaps came closest to being a major breakthrough in mathematics. This work was done entirely in India, before he left for England. Most mathematicians claim that Ramanujan's weak point was his work on number theory, despite his image of being mainly a number theorist.

visitors and friends and tried to enliven his company by a form of grim wit or black humour. But soon he had to face a new personal crisis, which also happened to be one of his last. It started with Komalatammal's efforts to send away Janaki to her parents' home. According to the mother's reading of the wife's horoscope, at least a temporary separation of the couple was necessary for the scientist to survive. The reading may have been a cover for the Indian folk wisdom which forbade sexual intercourse for those suffering from tuberculosis. But Ramanujan was, we have seen, a particularly sensitive person. He might not have been able to diagnose his mother's behaviour for what it perhaps also was: an attempt to express her latent jealous wish to possess the son all by herself, without feeling guilty about eliminating her rival. But he must have sensed, through the subtle communication that goes on in such close-knit relationships, that something more than his health was involved in the horoscope-reading: that his mother was trying to monopolize him by removing Janaki from the scene.

One does not know if there was any anger and if it triggered off the repressed anger that he had once turned against himself in the London Underground. After all, with his blind faith in astrology and expertise in palmistry, he may have seen in his mother's reading of the future the crueller hand of fate speaking through a temporal authority. He may even have felt guilty about resenting his mother's prescription. There is circumstantial evidence that Ramanujan's reaction was a mixture of all three. In any case, Komalatammal had not reckoned with her son's hidden strengths. This time, Ramanujan refused to send Janaki away to her parents.

Apparently, despite his adoration of his mother—even during his illness in England it was her that he mainly missed—and his sense of duty towards his brother Lakshmi Narasimhan, his patience with them was wearing thin. His wife at least feels that he wanted them to go away to Kumbakonam where his father, youngest brother, and grandfather were living. Janaki is a biased witness, but in this instance she may not be entirely wrong. Ramanujan always feared interpersonal conflicts and the chances of such conflict increased enormously with not only the two self-willed women but also the flamboyant Lakshmi Narasimhan living under the same roof. Moreover, Ramanujan's English period might not have made him a modern man, but it had exposed him to norms that legitimized some of his deeper inclinations. He was now freer with his wife in the presence of others, would constantly call her by ringing his bedside bell, and occasionally pull her towards him with the crook of the walking stick which he kept near his bed. He would even sometimes say that had he taken Janaki with him to England, he would not have contracted tuberculosis. In other ways too, he became less inhibited with his wife. He began teaching her the elements of science and using her as a secretary. Her job was to remind him of his various unfinished mathematical problems by using simple verbal tags which he provided her with.

In January 1920, Ramanujan was taken to Chetpeth, a suburb of Madras. His condition was deteriorating fast and it was felt that the air at Chetpeth would be beneficial. Others in the entourage were Janaki, his mother, grandmother, and Lakshmi Narasimhan. According to several accounts, it was now a withdrawn, depressed, and sullen Ramanujan who faced death. His black humour persisted, but the intense and penetrating look in his eyes betrayed the acute mental suffering and impotent rage of a man living under sentence of death, convinced that fate was against him. 'Namagiri sent me to England', he sometimes said, 'but she did not give me good health.' His anger, usually directed inwards, now became more outer-directed and free-floating. It set the emotional tone of his life and his strongest defences against it crumbled. Once in a while he would chew the thermometer put into his mouth or stealthily walk down to the kitchen and mix up or scatter the provisions on the floor. If Janaki stood at a distance, he would burst with anger and begin to throw things at her. He also often lost his temper with his mother and Lakshmi Narasimhan.

His last days were not happy. Poverty had dogged his steps all his life. Now he had money, but the culture of poverty stayed with him. He would get irritated with his brother for squandering his money or for being too stingy with it, and get agitated if he felt that he was not getting a proper account of the family's expenditure. Sometimes he said that, according to his own reading of his horoscope, if the family could somehow gift gold sovereigns equivalent to his weight, he would be cured; sometimes he assured Janaki that whether he lived or died, she would never again be short of money. The traditional image of tuberculosis did not help matters. It was called a *rajroga*, the disease of kings, and it required expensive treatment befitting its kingly nature; the emphasis was on rich food, change of air, and long stays in expensive sanatoria.

If Ramanujan was anxious about money, so was his family. Perhaps some of them were eager to make the best of the last few good days given to the family. While Ramanujan lay dying, they began to remove the few pieces of furniture and valuables from the house, anticipating his death. At one point, Janaki says, her brother had to call in the police to stop this.

Ramanujan's religious beliefs no longer provided complete protection against this collapsing world and the long, painful and, at times, ugly process of dying. One friend suggests that he even partly lost his faith in his favourite deities and considered them devils. His wife, too, feels that his faith in Namagiri wavered in his last days.[87]

Yet, one must hasten to add, Ramanujan simultaneously showed evidence of serene resignation and acceptance of 'fate'. As if, at one plane, he was still convinced that he should be tolerant of the force that had once given him success, but was now working against him. Until the end, he remained steadfast to his principles, showing the same single-minded, purposive unity of lifestyle that had characterized his whole career. As during his illness in England, he refused to take non-vegetarian protein food in spite of medical advice. His interest in psychic phenomena, astrology, precognition, and mystic mathematics was undiminished. And to the very end he remained an active mathematician—exploring, speculating, innovating. He would often work, against his doctors' advice, until one or two at night, trying to complete as many things as possible before he died. His free-floating aggression, which made him unwelcoming towards his friends in his last days, also helped him to snatch time to produce some of his finest creative work.[88]

Srinivasa Ramanujan died on 26 April 1920. He was then thirty-two, a ripe old age in a country which at the time had a life expectancy of less than thirty. Mathematicians too, Hardy affirms, are quite old at thirty. But not everyone is an Indian or a mathematician; in that case one might say he died rather young.

[87] It is possible that Hardy's faith in Ramanujan's rational agnosticism was partly the result of some comment made by the latter along these lines from his sick-bed. But in some systems of faith, anger against specific gods and goddesses does not signify loss of faith. After all, even in England, Ramanujan had tried to cure his tuberculosis with a *kavacha*.

[88] G. N. Watson, for instance, says (Valedictory address to the London Mathematical Society, quoted in Ram, *Srinivasa Ramanujan*, p. 72), 'Ramanujan's discovery of the mock-theta functions makes it obvious that his skill and ingenuity did not desert him at the oncoming of his untimely end. As much as any of his earlier works, the mock-theta functions are an achievement sufficient to cause his name to be held in lasting remembrance.'

V

Srinivasa Ramanujan did not live the life of a torn genius trying to reconcile science and culture, fighting the spectre of alienation, or desperately protecting a modern self against a traditional environment. His is the story of a conservative but integrated scientist, for whom ancient meanings and modern knowledge were one. I do not deny the psychological conflicts that dogged him throughout life, but they only marginally involved the content of his work. In fact, one marvels at the remarkably consistent way in which Ramanujan used mathematics to symbolize his inner states, without either damaging his mathematics or getting pre-occupied with the political or social implications of his success.[89] There were a number of reasons for this.

In the first place, Ramanujan's inner conflicts arose mainly from his attempts to cope with an unkind world outside, not with an alien self. He never internalized the Enlightenment culture of science;[90] traditions, whether expressed in strange customs or in mystic mathematics, were part of his innermost experience; he did not have to be apologetic about them. Nor did he seriously try to prove the modernity of his religious ideas, to Indianize western science, or to use his professional success to counter feelings of inadequacy.[91] Ramanujan was no conflicted proselytizer like Jagadis Chandra Bose, who tried to replicate in science the attempts of Swami Vivekananda and Shri Aurobindo to 'carry the message' of India to the West and that of the West to India. All these eastern Indians were sensitive to the West's political and cultural dominance, and their missionary zeal grew out of their sense of humiliation. Ramanujan was truer to Hindu orthodoxy; he neither sought any place in the metropolitan culture of knowledge nor showed any missionary passion.

This self-confidence was expressed in various ways. G. H. Hardy and J. E. Littlewood, when they said that Ramanujan had no

[89] Even Suresh Ram, in his otherwise naïve biography, recognizes that the 'majesty' of Ramanujan's search for self-definition 'lay in the harmony of his inner and outer beings', Ram, *Srinivasa Ramanujan*, p. 76.

[90] Ramanujan never said so, but seemed to draw a line between the content and the context of modern science. The former to him was part of the eternal verities; the latter peripheral to his concerns.

[91] He was not oblivious of the political and social changes taking place in India. But he kept his mathematics unburdened by his political or social beliefs. As if religion was the only load his work could carry.

clear concept of proof and showed little interest in its methodology, had in mind not merely the intellectual limitations of their Indian friend. They were also vaguely aware of Ramanujan's sturdy reliance on his own intuitive powers and insights.[92] For example, Ramanujan at Cambridge must have come to know that the days of formulae were more or less over and that his style of mathematics had become *passé*.[93] His friends in the university also tried to acquaint him with the new concerns of the discipline. But in success as in failure he remained true to his own version of mathematics. Though he made half-hearted attempts to adapt his style to the culture of modern mathematics, it was obvious to all who knew him that he could be neither easily taught nor formally educated. At most, he would learn on his own the things that interested him and were of value to him.

Ramanujan once reportedly said—and this would have deeply hurt his agnostic benefactor Hardy—'An equation has no meaning for me unless it expresses a thought of God.'[94] The statement, if his, was less a criterion of self-assessment than an acknowledgement of his God-given gifts. All his life he lived with a serene faith in his own supernatural precognitive powers and in the sacred origin of his mathematics. He was exposed early to his family's appraisal of him as a mystic genius, and his Indian friends mostly validated this estimate. His later exposure to western and Indian sceptics did not alter the self-image, which remained the traditional image of a *yogi*. He did not fancy himself to be an *acharya*, a Brahminic preceptor, but a man of superlative extra-sensory powers—a man possessed.[95]

There was also Ramanujan's strange love–hate relationship with the culture of modern science itself. Overtly, he did not fight the culture; he bypassed it. Covertly, he refused to collaborate with it.

[92] For instance, Hardy, *Ramanujan*, pp. 4–5. See the section on Ramanujan in Littlewood, *A Mathematician's Miscellany*, esp. pp. 86–88.

[93] See his letter to S. M. Subramanian, 7 January 1915, quoted in Srinivasan, *Ramanujan*, vol. 1, p. 21.

[94] R. Srinivasan, quoted in Ranganathan, *Ramanujan*, p. 88. Many western scientists have claimed that God is a mathematician; in the world of Ramanujan, mathematics was a system which integrated God, nature, and man.

[95] It must already have become obvious to clinically minded readers that Ramanujan's ego ideal was the mystic union with divinity often sought after by, apart from mystics, schizophrenic patients. This is significant in the context of his early object relations which also were, in many ways, consistent with the family dynamics of schizophrenics. However, more important was the manner in which this dynamic was integrated within the range of a particular, slightly esoteric, form of normality in his community.

This ambivalence deserves a digression on two differences between the modern western and the traditional Indian attitudes to science. First, a word on the more obvious of them.

The Indian tradition of science may be on the whole less positivist than its modern counterpart. But the tradition has been more open to certain empirical realities that Baconian science has come to acknowledge only reluctantly in recent years. Thus, many schools of Indian thought admit freely that science can be a product of a person's intuitive, infantile, non-rational self, and that the problems of the social responsibility of scientists can intrude into the very text of science. On the other hand, in Ramanujan's time, the operational principles of scientific creativity and the mainstream philosophy of science had become disjunctive. Creative science was often hypothetico-deductive, coloured by aesthetics and speculative thinking, and demanded personality resources compatible with these characteristics. The dominant philosophy of science, however, feared the cultural and psychological determination of science and favoured a crude form of inductive empiricism.

Hence, when exposed to western science for the first time as an adult, Ramanujan was caught between two powerful sets of attitudes. One agreed with what most scientists were saying but only a few were doing; the other agreed with what very few scientists were saying but at least the creative ones were doing. He therefore borrowed his philosophy of science from his inherited world-view to legitimize his work. Otherwise there was little in the mainstream philosophy to give meaning to his mathematics. (Threatened by the 'odd' way his Indian friend resolved this issue, Hardy re-read Ramanujan, as we know, as a self-consistent, agnostic English don. In his own case, the same manoeuvre did not help the troubled English mathematician overmuch. Facing the same contradiction, Hardy assumed the style of dissent of a modern European intellectual. Yet many of his 'saner' contemporaries promptly decided that he was an eccentric clown.)

Second, pure mathematics is unashamedly non-empirical, and the 'operative philosophy' of pure mathematics is comfortable with a rationalist attitude. With the growth of experimental science and empiricism, this rationalism was to some extent marginalized in the West. On the other hand, the Indian theories of knowledge, for whatever reason, maintained some openness to such rationalism, and at least, allowed some scepticism towards the positivist emphases on application, experimentation, control, prediction and testability in the 'real' world of the senses. Such openness might have crippled the hard sciences, but it certainly nurtured mathematical talents.

Naturally, Ramanujan's science did not try to be socially useful. Nor did it involve anything as non-Brahminic as experimentation, observation and proof. His was rather the clean, speculative non-dualism that has dominated Indian thought since about the eighth century. Two beliefs associated with this philosophy—the belief that contradictions represent aspects of the same indivisible truth, and the belief that 'true' knowledge would reveal the entire universe to be a unified living system—justified the mystical feelings of uterine, cosmic oneness frequently associated with scientific creativity, particularly with the *satori* experience of creative moments. Such feelings and such moments were not unknown to Ramanujan.

That is what gave Ramanujan the esoteric touch that seemed so attractive to the scientists of the inter-war years. Even his hard-headed collaborators could not remain immune to this charm. In his later years a slightly embarrassed Hardy rejected as unjustified sentimentalism the following paean to his friend's exotic appeal.

> It [Ramanujan's work] has not the simplicity and the inevitableness of the very greatest work; it would be greater if it were less strange. One gift it has which no one can deny, profound and invincible originality. He would probably have been a greater mathematician if he had been caught and tamed a little in his youth; he would have discovered more that was new, and that, no doubt, of greater importance. On the other hand, he would have been less of a Ramanujan, and more a European professor, and the loss might have been greater than the gain.[96]

However, as long as Ramanujan lived, Hardy had not recanted. It was the earlier, 'sentimental' Hardy who validated Ramanujan's self-identity and personal faith.

These issues of self-definition, autonomy, and creativity can be approached from another vantage point. That approach is summed up in two specific questions. Why did Ramanujan's orthodoxy

[96] G. H. Hardy, 'Notice'. It is possible to argue that the younger Hardy was more perceptive; the later Hardy was merely trying to make Ramanujan look a fully rational man, just as the latter's Indian admirers tried to make him look an occult magician.

never interfere with his creativity, while the modern idea systems like nationalism and Brahmoism could not check the intellectual disintegration of Jagadis Chandra Bose? Does the difference between the two men tell us something about the individual adaptations to the culture of modern science which individual scientists have attempted in India?

Seemingly, the answer to the first question is simple. Ramanujan died young; he could not grow into an entrenched symbol of Indian supremacy over the West or become a public attraction in his country. He did not get involved in organizational activities, remained outside the academic bureaucracy, and bypassed the ornate status hierarchy of the Indian educational system that engulfed many of his contemporaries. But all this, while obviously true, is only part of the story. One must look deeper for a full answer to this question. I shall try to do so by comparing his self-definition with that of Bose.

One difference between the two men was their area of study; the content of pure mathematics is patently less influenced by cultural and personal forces than most others and is less open to defensive projections. Its abstract, non-empirical structure is so formidable, and its ready-made delibidinized form so severe, that most cultural or personal themes lose their particularist edge when they enter the text of mathematics. It is literally the purest of sciences.

This purity has another source. Both modern physics and biology share with the social sciences the problem of 'contamination' from observation: the more detailed and intensive the study of a scientific phenomenon, the greater the likelihood of its being altered in the process of investigation. Particle physics, which among the hard sciences has most explicitly recognized this relationship between the observer and the observed, also suggests something else: that it is possible partly to transcend this indeterminacy only in the mathematical theory of a phenomenon.[97] In this respect, pure mathematics may be the closest to an observer-free science.

As a corollary there are certain specific demands which pure mathematics makes on the scientist's personality. Evidence has it that creative mathematicians and mathematical physicists have comparatively more impersonal identifications and tend to use withdrawal as a solution to their interpersonal problems.[98] In this respect, there is a fit

[97] Max Born, 'Man and the Atom', in Morton Grodzins and Eugene Rabinowitch (eds.), *The Atomic Age* (New York: Basic, 1963), pp. 590–601.

[98] For example, Roe, *The Making of a Scientist*; also D. C. McClelland, 'The

between the early environment that produces pure mathematicians and a Brahminic socialization. The latter, too, promotes a certain diffusion of identifications, encourages the use of the defences of isolation, denial, and intellectualization, and glorifies withdrawal from the profane world of real persons and events as a spiritual achievement.[99] Here is at least one reason why the traditional Indian ideal of mastery over self and the struggle to abstract oneself from worldly goods, instinctual needs, emotions, and social relationships, have found their supreme expression in mathematical creativity.[100]

This was one of Ramanujan's main advantages. Bose was working in what was then the unsure and inchoate disciplinary contexts of plant physiology and biophysics. He did not sense that while his vitalistic concept of science, rooted in the Upanishadic theory of life, could provide a valid philosophy of science, it could neither furnish ready-made scientific theories nor foreclose on alternative explanatory models.

Ramanujan's other advantage was his self-contained Brahminism which did not make overbearing demands on the content of his mathematics. Such demands were mainly confined to his personal life. The traditional Brahminic worldview had its own concept of science and it cared little for any other. Confident of itself, it also permitted a person to segment his life and maintain a certain distance from the non-traditional roles he might have had to take on for reasons of survival. Bose's semiticized Brahmoism could not permit this distance between what a person did and what he valued. Brahmoism was, after all, a rebellion against exactly this role diffusion and 'amoral' segmentation of life in India's modern sector.

In addition, the ideological intensity of Bengali high culture induced Bose to see the old and the new, the indigenous and the borrowed, as directly in opposition. The Bengali upper castes had always been aware of their peripheral status in India and they

---

Calculated Risk: An Aspect of Scientific Performance', in C. W. Taylor and F. Barron (eds.), *Scientific Creativity: Its Recognition and Development* (New York: Wiley, 1963).

[99] Spratt, *High Culture*; Carstairs, *The Twice Born*; Gardner Murphy, *In the Minds of Men* (New York: Basic, 1953).

[100] I have in mind here not only the mathematical achievements of ancient Indians, but also the comparative performances in contemporary India of, say, the mathematically-oriented scientist, on the one hand, and the applied scientist, on the other.

sought to consolidate their new salience, acquired through the colonial connection, by being belligerently Brahminic and modern. The efforts of the Bengali élite, the *bhadralok*, to define nationalism as essentially a modernist movement, rooted in rediscovered traditions, was a part of the same story.[101] The Bengali version of Indian nationalism was in more direct touch with modernity and was more aware of modern science. Its demands were personal *and* scientific, and it insisted on greater consistency between the private and the public. All these influences—Brahmoism, Bengali high culture, and nationalism—forced on Bose a more clear-cut decision between the traditional and the modern than on Ramanujan.

Finally, the absence of pronounced feelings of national and personal inadequacy allowed Ramanujan to use his fluid, unself-conscious, projective animism for the purpose of self-transcendence. Mathematics to him *was* a personal medium and he was very definitely concerned with his country's fate. But the former did not become the expression of the other. Bose's science, on the other hand, was coloured by the psychology of subjecthood and feelings of personal inadequacy. It allowed him neither enough autonomy nor flexibility. Unlike the European scientists of the age of faith and other Indian scientists more immersed in traditions, Bose in his science could not take advantage of the integrative strengths and range of his personality. Ramanujan was narrower in his politics and personal concerns but enjoyed the advantage of a less encumbered personality which had a wider range of fantasy life and symbolizing capacity.

This was unavoidable. For the very sensitivities that made Bose creative also made him sensitive to his colonial status. If his nationalism was a straitjacket, it was one he could not help wearing. One could even speculate that it was this sensitivity to issues of dominance and submission, and not any actual interference in the processes of research, which was often the main contribution of colonialism to intellectual decay. Ramanujan, less aggressively nationalistic and yet confident in his orthodoxy, escaped this double bind. And, if one could define intellectual autonomy as indifference to issues of dominance and submission, he was certainly the more autonomous of the two. Bose's modern

[101] For a brief discussion of some of the sociopolitical reasons of the sharper opposition, see Ashis Nandy, 'The Making and Unmaking of Political Cultures in India', *At the Edge of Psychology* (Delhi: Oxford University Press, 1979), pp. 47–69.

nationalism bound him too closely to the West, both in admiration and in hatred.

However, one could also define intellectual autonomy as a continuous search for new elements of identity which could be integrated within an indigenous frame, without humiliating the recipients, to allow the pursuit of a culture's distinctive version of universal knowledge. Perhaps at that plane, Bose, in his defeat, is a more relevant seeker of autonomy in the history of Indian science.

VI

A society survives by ensuring some consistency between the early developmental profile and the adult experiences of its members. In fact, its strength is this consistency in the life cycle of a large number of people. On the other hand, it is by stretching this consistency that a person gives meaning to whatever is new or disturbing to him and his society. If by using the symbolic repertoire of the culture, he succeeds in making this meaning authoritative or, at least, acceptable, he is recognized as creative. If he fails to do so, he is identified as an outsider.

So, when one discusses Ramanujan's response to his culture's definitions of authority, knowledge, inquiry and uniqueness, one must also refer to the new meaning he gave to science in his culture by stretching these definitions. This involved using the symbols and fantasies that were available to him as source materials for creative science in the twilight zone of his personality.

I have already said that the conflict between old and new in his society only marginally affected Ramanujan's developmental history. His adult experiences did sometimes contradict his early learning and selfhood. But the contradiction never deeply touched the core of his self. Perhaps the conflicts he suffered from did not require public resolution through ideological formalization. They demanded a re-enactment of early relationships through a more private rearrangement of his armoury of defences. Ramanujan's first authorities were a weak, passive, aloof father who threw into relief a possessive, overprotective, seductive mother. And he learnt early to cope with the passions these twin authorities aroused—by reifying, isolating, and systematizing the microcosm so as to be totally absorbed in the macrocosm. It was this crypto-schizoid dynamic that remained the fulcrum of his personality, in conformity and in defiance.[102] In his peak experiences he consistently used the symbols associated with motherhood and cognate ideas of unity, benevolence and magical power.

As in much of India, the experience of the mother as the most intimate authority figure, and of the father as a distant, intruding stranger were not unknown in Tamil culture. In it, critical sectors of life were presided over by female deities; and concepts of power, survival, nurture, and nature were inextricably linked with mothering. Much of the richness of one's fantasy life also derived from one's first object relation. And any access to one's deeper self or any regression at the service of the ego—as some psychoanalysts may like to describe the process—also led to the nuclear conflicts involving one's initial womanly authority. Creativity in such a world had to presume the capacity to exploit one's first projective identification with a cosmic maternal principle, one's feminine self, and one's concepts of defiance of a feminine authority and the reparation that had to be made for the defiance.

This play of authority, dissent and atonement was also a matter of psycho-ecological balance. In a culture where power seemed to reside outside the individual and nature often appeared as an absolute but fickle tyrant, the overlap between the individual's experiences of mothering and the community's experiences of nature's mothering was up to a point unavoidable. This overlap gave a different kind of sanction to the Indian man who sought to exploit his feminine self for creative purposes. It allowed him to defend himself against his fears of maternal vengefulness, and his basic distrust of the world of emotions and senses, through reified two-dimensional thinking.[103] It was mainly his mother's meta-science that gave meaning to Ramanujan's work. His 'insane vision' of creativity bypassed the world of modern science and mobilized his deepest and most androgynous self, for he had to master an animistic environment magically, to be in peace with and make

[102] Note that despite Ramanujan's pronounced obsessive traits, his personal experiences look remarkably similar to the family environment often found associated with schizophrenia in the West. See for example some of the papers in G. Handel (ed.), *The Psychosocial Interior of the Family* (London: Allen and Unwin, 1968). Yet, his family was not particularly atypical. This may be an indicator not of the society's obsessive or schizoid character, but of its ability to integrate certain obsessive and schizoid responses within its range of normality.

[103] Carstairs, *The Twice Born*, has dealt indirectly with this issue.

sense to himself. It is in this odd sense that Ramanujan's philosophy of science was unitary and he was one of the last representatives of the age of faith in science.[104]

[104] Interestingly, the same fantasies had parochialized Bose's science. For that matter, even the conflicts of Newton as he stood between the older magical science and the new science, and between two ages were, apart from the altogether different standards of creativity that he set, not very different. For a brief analysis of Newton from this point of view, see L. S. Feuer, *The Scientific, Intellectual and Sociological Origins of Modern Science* (New York: Basic, 1963).

Index